G000124839

Primary Maths
A Parent's Guide

Michelle Cornwell

Primary Maths: A Parent's Guide.
February 2008.
First Edition.
Published by M. Cornwell

Copyright ©2008 by Michelle Cornwell. All rights reserved

ISBN 978-0-9556920-0-0

Crown Copyright material is reproduced with the permission of the controller of HMSO
and the Queen's printer for Scotland.
All rights reserved. No part of this publication can be reproduced, stored on a retrieval
system or transmitted in any form or by any means, electronic, mechanical,
photocopying, recording or otherwise, without the prior permission of the author.
While every precaution has been taken in the preparation of this book, the publisher
assumes no responsibilities for errors or omissions, or for damages resulting from the
use of information contained herein.

Contents

Introduction

I have been a primary school teacher for the last 16 years and during this time I have had the opportunity to teach maths across the primary age range. I am a strong supporter of the National Numeracy Strategy as, having taught maths before and after its introduction, I can see the huge benefits it has to offer children. This book aims to address a need that, I feel, has existed since the introduction of the Numeracy Strategy. It offers parents a comprehensive explanation of the strategies children need to become numerate and presents the advice of The National Numeracy Strategy's 'Framework For Teaching Mathematics' (1999) and also the additional advice offered by 'The Primary Framework For Teaching Literacy and Mathematics'(2006) in what I hope is a readable, user friendly way. I hope to demonstrate to parents the strategies now used in schools to teach number and provide them with the understanding that they need in order to support their children appropriately.

What is the National Numeracy Strategy and when was it introduced into schools?

The National Numeracy Strategy was a new curriculum for mathematics, introduced into many British primary schools in 1999. Although it was not statutory (unlike the National Curriculum), it was expected that schools would want to take it on board as it offered a very comprehensive, specific guide for each year group in primary school. Teachers were given training through local education authorities and issued with a document called the 'Framework for Teaching Mathematics' which explained the new methods and strategies suggested for use from Year One up until Year Six. A separate section was also included which detailed expectations for the Reception year group.

Why was the National Numeracy Strategy needed?

The DFES publication 'Guide to your Professional Development Book 2: Effective Teaching and the Approach to Calculation' (May 1999) was issued as part of the professional development resources for teachers and it provided an outline of why the Numeracy Strategy was needed. Research from international studies and The Basic Skills Agency had shown that mathematical standards in British schools were lagging behind their European counterparts. In addition, both international and national research indicated that there was a need to 'improve children's and young adults' confidence and competence in mental calculation'. Teaching written methods too early, it seems, has an adverse effect on the development of mental methods and it is these mental methods that are required to use maths confidently in a range of everyday situations.

British schools have traditionally taught the vertical recording of 'sums' from quite an early age. For example, to work out the answer to 32+16, children would be expected to record it vertically as follows:

32
16+
—

Even extremely simple calculations would sometimes be presented in the vertical format. This early emphasis on written methods can encourage children to apply a procedure to numbers rather than to think about the numbers themselves. Vertical methods in the past have actively encouraged children to view numbers only as their single digits rather than looking logically at the whole number. For example, given the calculation 71-65, most children would have simply recorded the numbers (vertically) and followed the procedure (in this case 'borrowing' from the tens in order to subtract the units). If an error occurred, the child would often be unaware as they would have no real sense of the size of the numbers or any idea of what would be a sensible answer. As the two numbers involved are actually only 6 apart, it would be far quicker and more logical to work this out mentally. Looking at the whole numbers, rather than the single digits, leads children to gain a sense of the approximate size of the answer and therefore spot errors.

How has the National Numeracy Strategy improved teaching?

What the National Numeracy Strategy does is encourage children to gain a sense of the size of numbers and learn about the relationships between them. They are taught number facts and mental strategies which allow them to make use of what they already know to work mentally. Written 'jottings' (often in the form of numberlines) are made to help support and develop these mental methods and vertical methods of recording are now introduced much later, when mental strategies are already well developed.

'Expanded' methods of recording (which link to mental methods and develop understanding) are introduced as the first vertically presented written procedures. Only when these are fully understood are the standard methods introduced.

What are the difficulties posed for parents by the National Numeracy Strategy?

The National Numeracy Strategy has radically changed the approach to the teaching of mathematics and, although it has been very successful, it does pose a problem for many parents as they do not recognise the new methods and have no understanding of the significance of the strategies that are now taught. Maths is a subject that many children do worry about and, for parents unfamiliar with the new curriculum, supporting their child can be very difficult and confusing. Many of the methods taught prior to the introduction of The Numeracy Strategy actually conflict with the new methods and confuse children so it is important that parents know how to support their child appropriately.

How can parents gain an understanding of the National Numeracy Strategy?

As a teacher, I have often been asked for recommendations of books which explain the new methods but, although there are many practice books for children which demonstrate certain methods, there does not yet seem to be any means for parents to gain a comprehensive overview of how number is now taught. Some parents do want to see and understand 'the big picture'. The aim of this book is to provide interested parents with a guide that explains the strategies and methods used in each year group and gives them information about the types of calculation that their child should be able to solve using these strategies.

Can this book be used purely for reference or does it need to be read sequentially?

It may be that your child does not require a great deal of support but you feel that you require a reference book for when your child brings home homework that uses new methods with which you are unfamiliar. If this is the case, it should be fairly straightforward to simply look up the strategy or method in the appropriate year group. It is worth noting, however, that occasionally the strategy you are looking for may be actually in the year above or below. Sometimes teachers refer to the curriculum for the year either side of the one they teach to either consolidate or extend understanding of strategies. If this is the case with your child, you would obviously work from the example which matches your child's homework, regardless of the year group in which it appears.

Although this book is designed to be used simply for reference, if you have a child who struggles with maths, I would recommend reading the whole chapter which pertains to their year group and also the chapter before (pertaining to the previous year group). Many of the earlier strategies are built upon in later years and it can be useful to see earlier stages of a strategy to help to understand its significance. If your child has one strategy that they do not seem to have grasped, you are then in a position to start it at a simpler level in order to revise concepts that may not be secure, and then build upon them as necessary. It is sometimes the case that a child fails to understand a concept because some very simple aspect of it has somehow been missed at an early stage and the child has therefore not had a secure base to build upon.

Once these chapters have been read through, you will begin to gain an understanding of how strategies link together and build upon different areas of knowledge. You can then look at the chapter which pertains to your child's year group and refer to this in more detail to support your child in areas in which you feel they may need it.

Does this book cover the entire curriculum for mathematics?

No. Many areas, such as shape, symmetry, data handling and problem solving (to name but a few) have been omitted because I wanted to focus on the many aspects of number and how they link together. Addition, subtraction, multiplication and division are covered comprehensively and the main strategies used to teach these are explained and de-mystified. My aim in writing this book was to provide parents with an understanding of the new methods (both mental and written) for teaching number. Bear in mind, however, that the ability to use and apply mathematical skills and knowledge is essential so it is important that, as soon as your child has a certain skill, you encourage them to use it to solve everyday problems.

Will I understand the new mathematical terms?

The National Numeracy Strategy uses a significant amount of very specific mathematical vocabulary with which parents are likely to be unfamiliar. There is, however, a glossary at the back of the book which gives an explanation of the key vocabulary. This may need to be referred to quite frequently until the new terms are understood.

What is the revised numeracy framework?

In October 2006, a revised version of The Framework for Teaching Mathematics was published entitled 'The Primary Framework for Teaching Literacy and Mathematics'. (Throughout this book, I have referred to it as the revised framework). This raises the expectations for children in some areas of mathematics and provides greater cohesion between the objectives for the Foundation Stage and Key Stage One. However, the fundamental approach has remained the same.

Because it is still relatively new and, at the time of writing, resources and training are still being delivered to schools, it is difficult to be certain that every aspect of the revised framework concerning number has been included in this book. Although the Numeracy Framework has not changed radically, the revised strategy has presented *some* additions and changes to the original framework in terms of what should be expected in each year group. I have consulted the materials from the revised framework when including advice for each year but, where I have been unable to find specific advice about a certain strategy, I have assumed it to be unchanged and given examples from the original framework. I have also assigned some written methods to particular year groups whereas in the revised strategy they have been presented in developmental stages. Please bear in mind, therefore, that the year group examples are only a guide. The year group content will vary from school to school and from child to child and, if you find that the examples in your child's year group do not fit their experience, it would be sensible to look to either the chapter above or below in order to find the developmental level of your child.

Chapter 1 :   Numberlines

Numberlines are central to the teaching of the National Numeracy Strategy, especially in the early years of primary school. They are important as they encourage children to visualise numbers and give them a flexible way to represent their thought processes. However, they are completely new to the majority of parents and their purpose can be unclear if they are not understood and used appropriately. It is for this reason that I have decided to devote the initial chapter of this book purely to the use of numberlines. I hope that, by demonstrating the basic ways in which they are used right at the start, confusion will be avoided later when they are encountered as a part of more complex calculations.

Number tracks have been used in Nursery and Reception classes for many years. Often they are painted on the floors in playgrounds and used as part of outdoor activities. Number tracks can also be printed on boards or in books and used for a variety of counting activities. The numberline simply builds on the idea of the number track, but instead of the numbers being placed within squares as on the number track below

| 1 | 2 | 3 | 4 | 5 | 6 | 7 | 8 | 9 | 10 |

Figure 1.1

they are printed above or below divisions on a line of numbers (see Figure 1.2).

Figure 1.2

In the early years, the numberlines used usually start at 0 or 1 and end at 10 or 20. Obviously, as the children become more familiar with larger numbers, a wider range of numberlines can be used. Numberlines beginning at 0 or 1 and stretching to 100 and beyond will help children to become familiar with the positioning of larger numbers.

Numberlines to 100 usually have only the multiples of ten marked upon them, although they often have the actual calibrations for the numbers in between (see Figure 1.3).

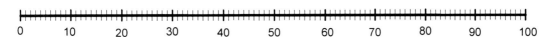

Figure 1.3

Teachers initially use numberlines to consolidate understanding of the order of numbers and children may count forwards and backwards along number lines to become familiar with number order. They may also be asked to position missing numbers on numberlines where some or all of the numbers are hidden. They are taught that numberlines do not have to begin with 0; they can begin and end with any number. Children are often given just one part of a numberline to investigate. For example, in Year One they may be asked to find the number to which the arrow is pointing in the diagram below.

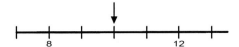

Figure 1.4

In Year Two the same question may be asked but with a slightly more difficult diagram.

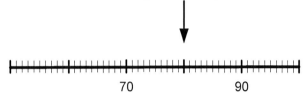

Figure 1.5

Once children can order and position numbers on numberlines, they can begin to use them to add and subtract.

In the early years, children begin to understand addition as counting on and subtraction as counting back (see Chapter Three for a more detailed explanation of this). Number tracks and numberlines are very useful for developing this understanding. The transition from number track to numberline is usually made in Year One when children are encouraged to add or subtract by drawing jumps on a printed numberline. For example, to solve the calculation 8+3, children would initially be encouraged to count on by jumping their finger along the numberline or track. So in this case, they would start at 8 and jump along 3 more, counting (1, 2, 3) as they do so. The number they would land on is 11 so 8+3=11. Once children are familiar with this process, they can begin to record it on printed numberlines by drawing the 3 jumps as they make them. They may begin by drawing on 8 jumps, then 3 more as shown below.

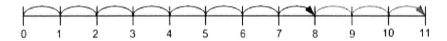

Figure 1.6

However, if they are using numberlines regularly, it should soon become obvious that this is unnecessary. (If it doesn't, even after it has been pointed out, then they obviously need more practical work using objects to add and subtract as they do not have what is required to understand addition as counting on. See Chapter Three).
Soon, they should be able to work out the addition 8+3=11 as shown below.

Figure 1.7

The same process can be carried out with subtraction, only the children obviously count backwards instead of forwards.

By giving children frequent experience of using printed numberlines to solve addition and subtraction calculations, they should eventually become confident enough with the order and position of numbers to draw their own blank numberlines and visualise where the numbers are on them.

To give an example of drawing a blank numberline, imagine a child is given the calculation 23+6. Some children would be able to count on 6 more from 23 using their fingers. However, an alternative method is to simply draw a blank numberline (which is basically just a horizontal line) and imagine roughly where the relevant numbers are, in this case 23. They can then draw on the numbers and the jumps just as they would with a numbered numberline and count what they have drawn. So first they would draw:

Figure 1.8

Then count on 6 by drawing 6 jumps (as 6 is to be added) like this:

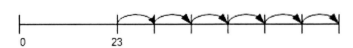

Figure 1.9

It is worth noting that only the relevant parts of the numberline need be drawn so, if they are very confident, they could draw:

Figure 1.10

They would then count on 6 jumps in exactly the same way.

Figure 1.11

Finally the children count on from 23, using their jumps (24, 25, 26, 27, 28, 29) and write the number they reach at the end.

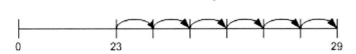

Figure 1.12

Because they reflect the mental process, children can use numberlines at their own level to tackle a problem. For example when working with a calculation such as 89-25, children who are developing their understanding of counting back in tens and ones may draw a numberline as follows:

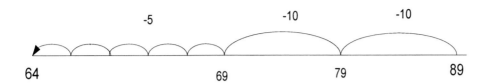

Figure 1.13

However, if their mental methods are more secure, they may be able to solve it without splitting the 20 into 2 tens or counting back in ones. If this is the case, the numberline would be recorded as follows:

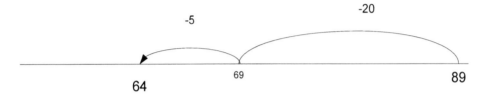

Figure 1.14

This aid to the mental process becomes invaluable when working with larger numbers. For example, if working out the calculation 456+327, a blank numberline can be used to count on the hundreds, then the tens, then the ones (see Figure 5.11).

Figure 1.15

Numberlines are also used for multiplication and division and again can be used very flexibly. Not every number on a numberline will need to be drawn. The child need only write those that are relevant or that they need to help them. For example, they may record 5x3 in one of the ways shown below, depending on how useful it is for them to write on the multiples of 5 as they count.

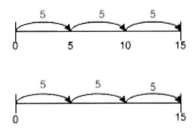

Figure 1.16

It doesn't matter which way they record it as long as it helps them to work out the answer. Similarly, with division, if asked to solve 30 divided by 5, as long as they are confident with the concept of grouping (which is explained under the relevant sections on division), they can work out 'How many 5s are in 30?' as follows:

Figure 1.17

Although this does not seem any easier than using other methods (such as counting in fives on their fingers), it becomes very useful later on when larger numbers are needed such as, 'How many 15s are in 75?'

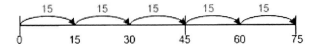

Figure 1.18

Using numberlines for division is also useful as it can help children to understand the repeated subtraction that grouping involves (see explanations of division in later chapters) and allows them to visualise the relationship between multiplication and division.

One of the great advantages that numberlines offer is that they are incredibly flexible and because of this can be used with numbers of any size. They can be used in the later years of primary school to help children to order and position 3 and 4 digit numbers, decimals, fractions and negative numbers, as well as being useful for multiplication and division.

To summarise, numberlines are important as they allow children to represent their thought processes as they are working out. They can be used throughout the primary age range in conjunction with other strategies to develop understanding of numbers. Because they relate so closely to mental processes, they can lead readily to the development of mental methods so that eventually calculations can be solved without the visual representation of the numberline.

Chapter 2 : Reception

To be meaningful, maths needs to be taught in context and this is particularly the case with very young children who find abstract concepts difficult. The majority of activities suggested in this chapter therefore are unlikely to be new to many of you who already incorporate number into everyday activities with your child. I am hoping, however, that my ideas will reassure you of the value of what you already do and explain how to take your child further if you feel your child is ready for more of a challenge.

The work carried out in the Reception class is based around gaining a basic understanding of numbers. This involves developing two aspects of counting: the ability to count objects and the ability to rote count (or chant numbers in order). Children will also gain an appreciation of the relationships between numbers and gain an understanding of addition, subtraction and even early multiplication and division, through practical play based activities. It is relatively easy for parents of young children to consolidate what goes on in the Reception classroom as so much of it is rooted in play and everyday experiences.

Rote Counting

To develop an understanding of the relationships between numbers, children need to develop their awareness of number order through both songs and rhymes and through rote counting. Rote counting simply means chanting the numbers in sequence: 1,2,3,4 etc. Regular singing of number rhymes and songs are of vital importance as they help to create familiarity with the names of numbers and number order. It is a good idea to make sure your child has regular practice with both songs that involve counting forwards (e.g. 'This Old Man...', 'One Man Went to Mow..' etc) and songs which involve counting backwards (e.g. 'Five Currant Buns...', 'Ten Green Bottles...' etc). Children can be made more aware of mathematical relationships, language and concepts through appropriate questions during these songs and rhymes. For example, ask your child to show fingers to represent numbers and ask them to make predictions at the end of each verse. By asking questions such as 'So there were 3 and one more will be ...?' or 'So 3 add 1 is...?' or 'The number before 6 is ...?' you can help build up confidence with mathematical language as well as with the concepts of more, less, before and after. When children have lots of practice with this type of activity they begin to develop their understanding of the relationships between numbers, telling you that one more than 3 is 4 without counting.

Counting Objects

It is important that children count objects regularly as part of everyday activities. Activities do not need to be especially set up as opportunities for simple counting exist in most everyday contexts. For example, simple activities such as counting each step as you climb the stairs, each coat button as it is fastened or counting how many plates are needed to set the table will help build familiarity with the language of number (such as enough, too many, the same, more, less, altogether etc) and also help your child to understand the concept of one to one correspondence which is vital for early number work. This basically means that, when counting, the number that is said corresponds with the correct number of objects. It is common for some children to learn to rote count confidently to 10 or beyond but be unable to count objects accurately because they do not yet have one to one correspondence. This can only really be developed through practice. Encourage your child to touch or move objects as they count them as this helps to reinforce the idea that each number name represents each object that is counted. They then gradually come to realise that the last number counted indicates the number in the set and that whatever order the objects are counted in, the number in the set will remain the same.

Often children will have one to one correspondence with up to 3 or 4 objects in a group but then lose it with larger groups of objects. If this is the case, simply count smaller groups of objects until your child gains one to one correspondence with these then gradually increase the size of the group. Although in Reception children are expected to reliably count groups of initially up to 5 objects, then up to 10 objects, it is important that your child also regularly counts smaller amounts even if you feel it is too easy for them. This is because it helps to build their confidence and gives them a sense of the size of numbers.

It is worth mentioning that small children may lack understanding of many of the concepts about counting that we take for granted so it is also important to count a variety of objects in a variety of settings in order to consolidate their understanding. For example, make sure that objects to be counted sometimes vary in size or shape and also in arrangement. Many children can confidently count objects arranged in a row but are unable to count randomly arranged objects as they cannot remember which ones they have counted. If this is the case with your child, you can develop their understanding by discussing how to make counting easier, perhaps by moving the objects as they are counted or arranging them in a different way so that they can be counted more easily. Even the spacing of objects can confuse some children and they may believe that there are more there when they are spread out to cover a wider area. As they gain experience of counting, they gradually come to realise that the number in a set does not change if nothing is added or taken away. This concept is called 'conservation of number'.

Once your child becomes very confident when counting objects they can be encouraged to count by looking but without touching. As well as counting groups of objects, children can be encouraged to count pictures, actions, sounds and rhythms. When counting sounds or rhythms it may help to have the eyes shut (some children find it easier with their eyes shut as it removes visual distractions). For example, count the number of pennies dropped into their money box by listening to them fall or count how many times they can hop, clap etc.

If your child particularly enjoys outdoor play then make a point of playing outdoor games which involve counting, such as skittles, hoopla, throwing balls or beanbags into a bucket or counting the number of hops or skips that can be done before the timer runs out. A variation of 'hide and seek' is a particularly popular game; a given number of objects (5 is good to start with) can be hidden for your child to find. They will be required to count them as they find them to work out how many more are still hidden and to work out when they have found them all. These types of activities are enjoyable and give children a purpose for counting. They also incorporate addition and subtraction so questions that require your child to understand and work out what has happened can help develop their understanding of these concepts. For example, 'You had five skittles but three were knocked over, how many were left standing? If only one was knocked over, how many would be standing?' 'You threw two beanbags in the bucket, then two more. How many were in the bucket altogether?' You can encourage your child to use their fingers to help them work out the answers to hypothetical questions; this can help them to become confident with the concepts of adding and taking away.

Eventually such practical experiences may give children the confidence to recognise some smaller amounts without counting. Games involving the use of dominoes or a dice are especially useful for this as children come to recognise the arrangement of the spots. Using fingers to show amounts is another way to help children to begin to recognise amounts without counting. However, it is important to remember that just because a child can recognise a certain amount just by looking at it does not mean that s/he has one to one correspondence and can count that amount of objects. Sometimes, the development of mathematical concepts is not linear and they may not be grasped in a particular order.

More and Less

As well as counting objects, it is a good idea to compare the sizes of sets of objects using language such as more, less, most, least etc so that your child can gain an idea of the size of groups in relation to each other and an understanding of more and less. When they are playing, make staircase patterns (adding one more each step as shown in the diagram 2.1

below) using pegs and pegboards, bricks, Lego etc to help your child to understand the relationship of one more or one less and see the pattern made by repeatedly adding or subtracting one. This is important as it is the beginning of mental methods for addition and subtraction which can then be built upon. If your child knows instantly what one more/less than a number is, it indicates that they have a secure grasp of the order (i.e. pattern) of the numbers. Some children also begin to use this knowledge to begin to work out what two more/less would be. However, be warned: the idea of one more/one less may well be understood in practical contexts but it actually takes a great deal of practice in a range of contexts to consolidate this understanding so that your child can answer the abstract question 'What is one more/ less than ...?'

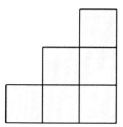

Figure 2.1

A staircase pattern, showing the effect of adding one more each step.

Estimation and Approximation

Encourage your child to estimate amounts of objects around them at times during daily activities as this will help to develop their understanding of the size of a group and is the beginning of approximation. In later years, children will be expected to approximate then check their work as a matter of course so it is a good habit to get into. Approximating can be done incidentally with comments such as 'I wonder how many are there? How many do you think?' Once they have had a guess, they can check by counting. In this way, their estimates will gradually become more accurate. Children often make ridiculous estimates to begin with simply because they have no real understanding of the relative sizes or values of numbers. Their guess therefore will not seem ridiculous to them. As your child's experience grows, their estimates will become more realistic and discussion can begin to guide them to a reasonable answer. By asking questions such as 'Do you think it is more than 3?' 'Is it less than 10?' you can help them to think about numbers in relation to other numbers and they can use these comparisons to help them. During such activities it is wise for you also to guess and not always to gain exactly the right answer so that your child does not have a fear of being wrong and sees that sometimes there is a place for approximation.

Counting Forwards and Backwards

Counting forwards is frequently done, both at school and at home, and this is why children become very good at it. Counting backwards tends to be practised far less frequently. However, it is important for two reasons: firstly it helps children to develop their understanding of the relationship between numbers and secondly it is an important skill, needed for solving many subtraction problems. It is therefore important to help your child to count backwards as well as forwards.

In Reception, children will initially need to practise counting forwards and backwards to and from 5 and then 10. Games involving rockets and blast off are a good way to do this. Children seem to simply love chanting backwards from 10 and shouting 'Blast off!' at the end! Once children can count forwards and backwards to 10, they will need to practise counting to 20 then beyond. They commonly make mistakes such as twenty eight, twenty nine, twenty ten... so lots of practice crossing the tens boundaries is needed. Puppets and toys can be used to demonstrate incorrect counting for your child to correct. Even when they cannot consistently count correctly themselves, they often can spot incorrect counting by others. Errors such as counting in the wrong order (1, 2, 4, 3, 5) or missing numbers out (1, 2, 3, 5, 6) or repeating numbers (1, 2, 3, 4, 5, 5, 6) can be made by the puppet and your child can gain confidence and develop their understanding by spotting the mistake and saying what is wrong.

As rote counting becomes more automatic, children can be asked to continue a count started by someone else or count from different numbers instead of always starting with 0 or 1. One way to do this is to pretend that a toy keeps forgetting its numbers and needs somebody to help it finish off counting. The toy can then count to 4, for example, then stop and your child can carry on. Another idea to try is to ask your child to count from one number to another. For example, start counting at 5 and count on until 7 or begin at 6 then count 3 more numbers (using fingers). As counting backwards is improved, the same types of activities can be carried out by counting back, e.g. start at 9 and count back to 4 or start at 8 and count back 3 etc. Although they are quite challenging for some children at first, these types of activities can be useful as, not only are they the beginning of addition and subtraction in the form of counting on and counting back, they also encourage your child to hold a number in their head whilst counting, a skill needed for later calculations. By developing their confidence with numbers in this way, they are laying the foundations for later mental methods.

Number Recognition and Order

Initially much of the work in Reception is based around counting in different contexts and children are encouraged to record numbers in their own way at first through play activities. This may mean drawing 2 dots to represent 2 objects found in 'hide and seek'. This type of activity can then lead to discussion about the need to record numbers sometimes and the numerals can be introduced. Number labels could also be made as part of games for certain toy boxes or number plates made for their bike etc. so that your child can practise writing numbers in informal contexts. Young children need to understand the significance of numbers in everyday life so make sure you look for and point out numbers in their play environment. For example, draw attention to the numbers on clock faces, on phones, calculators etc. Numbers can be made when playing with the playdough or finger painting and guessing games can be played when the child has to guess which number is being written in the air or on their back. There are also a number of commercially available jigsaws and games which are based around number recognition.

Number Relationships

It is important for children not only to recognise numbers but also to gain an understanding of numbers (and the relationships between them) up to 10. Practical activities and songs and rhymes will have started to develop their sense of the size of numbers by talking about more/less, bigger/smaller etc. Encourage your child also to match numerals up to 10 to sets of objects and collect sets of objects to correspond to given numerals. Often children can count a group of objects that are laid out for them but are unable to collect a given number of objects from a larger set. (This is a harder skill).Children commonly collect too many and then put them all back and start again. You can help your child to develop their understanding of number relationships by asking appropriate questions. For example, 'Oh dear. You've got too many. Do you think you should add some more or should you put some back?' These sorts of questions will help your child to think about and eventually come to understand the relative sizes of numbers.

Number jigsaws are useful for helping your child to place in order numbers to 10; you can develop their understanding further by asking questions requiring them to say which numbers are bigger (greater) or smaller than others. Encourage them to think about which number is before or after (or one more/less) than another up to 10. You could also ask them to say which numbers are in between others, e.g. 'Which number comes in between 6 and 8?' Ordering random numbers, such as 7, 4 and 9, from the smallest to the largest or vice versa is a useful way to find out if your child understands the sizes of numbers in relation to each other. Children often enjoy it when a number from a puzzle is hidden and they have to guess it.

Once your child confidently recognises numerals, they can also begin to count along a number track. In school a simple number track may be made by giving children number cards and arranging the children themselves in order. A large number track or numbered stepping stones are also useful as they allow children to jump along a track and count numbers as they jump. Mathematical language such as 'start on number 3 and add 2 more' can be used as part of games and physical activities. Children can also be encouraged to predict before counting which number they think they will land on so that gradually they will begin to see the relationships between numbers and work out some answers mentally. The pattern between one more/one less is often the first to be worked out in this way.

Games can be played with homemade number tracks, giving your child points for each right answer and encouraging them to ask you questions and give you points for correct answers. I have included a chapter at the end called 'Games to Play' which gives ideas for games to develop an understanding of number relationships. However, remember these games should be fun and last minutes rather than hours. Young children learn far more quickly in short, sharp bursts than in long, intense sessions!

Counting in Tens

Once children have a good understanding of numbers to ten and are able to count well beyond 20, they can begin to learn to rote count in tens. Children in Reception need to gain confidence when counting forwards and backwards in tens e.g. 10, 20, 30....to 100 and should begin to say the number that is ten more or ten less within this sequence. Obviously, the more they practise, the easier this becomes.

A hundred square[1] will be used in school and wallcharts displaying hundred squares can be bought from many bookshops. It is a good idea to have one so that your child can actually see the numbers that they are counting, the pattern they make and the relationships between them.

If your child is very confident, they may be able to count forwards or backwards in tens from different starting points (e.g. 40...50, 60, 70...) and continue a count started by someone else. Counting forwards and backwards in tens is an important part of developing understanding of place value (tens and units; see glossary for more detailed explanation). Again you may want to challenge your child by giving numbers at which to start and stop counting, e.g. 'Count back from 50 to 30. How many did you count?' (Say 50, then show finger as count 40, show another finger as count 30. Two fingers are shown because 2 tens were counted, i.e. 20). As with the activities for counting in ones, toys and puppets that make mistakes or need help are a good way to include these types of activities in everyday play.

Rapid Recall

Although children in Reception are not required to have mental recall, repeated practical work does often allow some children to have instant recall of some facts. This can then be built upon in later years. It will be useful for your child to be able to instantly say the number before or after another without counting (up to 10), to become confident, through practical experience, with the addition and subtraction facts within 5 (for example, 2+3=5, 5-3=2 etc), and to begin to know doubles up to 5+5 by heart. Instant recall of these facts can be gained through repetition of counting activities, games and songs.

---

[1] See glossary and also Chapter 4, page 40.

Mental Calculation Strategies

Although most of their work is practical and children in Reception are not expected to work mentally, many of their activities are laying the foundations for mental work. For example, counting forwards and backwards from different numbers is one of the skills needed for mental addition and subtraction; counting in twos, fives and tens is laying the foundations for the understanding of multiplication. Activities such as these will be extended and consolidated in later years in order to build a secure understanding of the number system.

Addition

By the end of the Reception year, it is hoped that, through their practical experience, children will be starting to understand:
- That addition makes numbers bigger (except when adding 0).
- That addition can be done in any order (i.e. that 4+5 is the same as 5+4).
- That addition can mean combining two or more sets.
- Some children will also begin to gain an appreciation of addition as counting on through practical work using board games, stepping stones etc.

Addition and the associated language is experienced mainly through play activities, although fingers or objects may be used to solve certain hypothetical problems.

Combining Sets.

Addition needs to be understood as combining two or more sets. This involves counting each group then counting the total. Again, this is easily experienced through play. For example, 'I have 2 boats in the water and my friend has 3 boats in the water. How many boats are there in the water altogether?' There are 3 and another 2 so that is the same as 5. When children can confidently add 2 groups then 3 groups can be added. During the Reception year, the teacher will model the recording of addition and also encourage children to repeat the number statement e.g. 3+2=5. (They will also demonstrate the number sentence on a number track in order to develop children's familiarity with the concept of counting on). Children are not expected to formally record addition until they have a very secure understanding and I would advise against your child doing a great deal of formal recording at home. If you feel that they are very secure with their number facts, then you can challenge them by asking them to apply their knowledge to various problems such as asking questions or setting problems that involve the addition of three groups instead of two, using slightly larger numbers or using addition and subtraction statements to help them see the link between the two. For example, 'I went to the shop and bought 5 sweets then my friend gave me 3 more. How many sweets do I have?' 'Then I ate the 3 sweets my friend gave me. Now how many do I have?' Talk about how, if you add a number then take the same number away again, you will be left with the original amount.

To help your child develop an understanding of number bonds (explained in glossary), groups of objects should also be split in different ways. So, 'How many ways can I share these 5 pencils between 2 pots?' 'How many different ways can you show me 6 using the fingers of 2 hands?'

At home, activities such as these can easily be included in play. For example, 'Teddy has hidden 4 cars in the garage. Some are red and some are blue. Can you guess how many of each he has hidden?'

Counting On

Once children are confident with the concept of addition as combining sets, they can begin to develop their understanding of addition as counting on. This can be done through activities such as putting a certain number of objects in a line, for example 5, then counting on from 5.

E.g. 5 socks on a washing line, add 2 more so the total is 5...6, 7. It can also be demonstrated using fingers, e.g. hold up 5 fingers then count 2 more...6, 7.

When playing board games, attention can be drawn to the fact that you are counting on from the last number. If your child finds this difficult initially, plenty of practical work should be done with giant stepping stones or number tracks drawn with chalk outside to consolidate their understanding.

The next step is to count on from a number when it can't be seen. So, for example, 5 add 2 could be done with 5 fingers behind the back then count on 2 more or 5 sweets could be hidden in a bag then 2 more added. These types of activities will be understood more readily if they are part of everyday activities. For example, 'There were 5 pennies in my purse. I find 2 more on the window sill, so how many do I have now?'

In Year One your child will develop their understanding of addition as counting on using a number track or a numberline so, once your child is very confident with addition as combining groups, you could demonstrate addition as combining groups on a number track by placing 3 red cubes, for example, on numbers 1 to 3 then 3 blue cubes on numbers 4, to 6 on a number track such as the one below and counting the total. As you do this, you can begin to point out that you do not actually need to count the first 3 cubes because we know there are 3. This will help your child to begin to make the transition from counting on practically to counting on using a number track.

1	2	3	4	5	6

Figure 2.2

Doubles and Halves.

Another very important part of the Numeracy Strategy is doubles and knowledge of doubles (and halves) can really help with mental calculations in later years. Attention can be drawn to doubles in play activities, such as 'There are 4 wheels on the red car and 4 wheels on the yellow car, so double 4 will be ...?' Later, hypothetical questions can be posed and the answer can be worked out either by combining groups or by counting on depending on the level of the child and their experience. Using their fingers to show doubles (e.g. 3 fingers on each hand to show double 3) is a good idea as it is something they can easily refer back to when asked a hypothetical question or posed a problem.

Subtraction

By the end of the Reception year, it is hoped that children will have started to understand that subtraction:

- Makes numbers smaller (except when subtracting 0).
- Can be seen as taking away an amount from a set of objects.

They should also be familiar with finding the difference (by counting up) in practical contexts.

Subtraction should be experienced as taking away objects from a larger set. For example, 'There were 6 trains but if we tidy up 3 how many will be left?'
E.g. 6-3

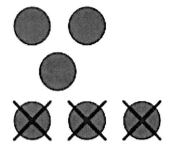

Figure 2.3

Or

Figure 2.4

It can be shown as taking away on a number track by crossing out the numbers near the end.

Figure 2.5

Subtraction can also be solved by finding the difference in the context of 'how many more'.

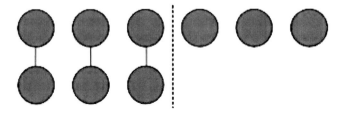

Figure 2.6

For example, 'I have 3 sweets but my friend has 6 sweets. How many more sweets does my friend have?'

Although 'how many more' involves counting on, it is actually a subtraction as it involves finding the difference between two numbers. Even at this young age, children can begin to understand it in a problem solving context. This concept can be developed when your child is playing. For example, when playing with a toy farm, a rule could be made by the farmer that only five cows are allowed in each field. You could then, for example, put three in and ask your child to work out how many more will be allowed. Initially your child may need to add the cows physically. If this is the case, you can help them to develop their understanding of number by allowing them to put in the number they think is correct then counting them together. If there are too many or not enough, then discuss what you need to do next: do you need to add some more or take some out in order to have five in the field? These types of games help children to understand the relationships between numbers and their relative sizes and can really improve their confidence. With plenty of experience of these types of activities, children can sometimes start to predict how many more by counting up in their head or by using their fingers. This can then be extended by asking your child to hide their

22

eyes while you remove some cows from a field of five and ask them to predict how many more will be needed to total five again.

Multiplication and Division.

It may be a surprise to see multiplication and division included in the Reception curriculum. However, play activities offer an ideal opportunity to use multiplication and division in context. Working with pairs and counting repeated sets with the same number will obviously help children to develop an understanding of the concept of multiplication. Children should gain an understanding of pairs very early on through their play activities. Socks, gloves etc can be placed in pairs in role play activities and matching games involving pairs can also develop this understanding. There are many commercially available games based around matching pairs and these are useful, not only for understanding the concept of pairs, but also for developing the visual memory, an important part of reading.

Once children have a clear idea of what pairs are and why some things are grouped in this way, they should begin to see the relevance of counting in twos as paired objects are counted. As with counting in ones, rote counting is useful to help children remember the sequence 2, 4, 6, 8 etc. Activities which begin to introduce the pattern of odd and even numbers are also introduced. Counting in ones but whispering every other number so that all the odd numbers are whispered (or all the even numbers) is a good idea as is colouring patterns of odds and evens on number tracks so, for example, all the odds could be red and the evens blue. Once the numbers are coloured, your child can be asked to close their eyes and tell you the colour of a certain number. This helps them to visualise numbers, which helps develop understanding and memory. Certain number rhymes, e.g. '10 Fat Sausages', '1, 2 Buckle My Shoe..', can also help consolidate understanding. Counting in tens can be introduced as a chant and used to count fingers on hands, toes on feet etc.

Practical problems which involve counting repeated groups of the same size or sharing objects are a good idea. For example, 'Each teddy has 2 ears, so how many ears would 5 teddies have?' Sharing out objects such as cards or dominoes between players of a game or grouping biscuits so that the teddies all have the same amount helps to give an understanding of the idea of division as sharing and grouping so that when it is encountered formally it is more likely to be grasped confidently. Learning to count in twos and tens also forms the basis for later work involving multiplication and division.

In summary, the Reception curriculum is very similar in many ways to the pre-Numeracy Strategy curriculum. It is still very much rooted in play activities and practical work and attention to number in everyday activities will help your child to see the relevance of numbers and counting. There is, however, a more structured approach to the teaching of number relationships. You can help support this approach by using number tracks, number puzzles and simple number games to help your child to visualise these relationships. Using board games, in particular, can help your child to develop an appreciation of counting forwards and backwards in preparation for later work on numberlines. By practising these types of activities, in short bursts, you can help your child to lay firm foundations for mental skills which can be built upon in later years.

Chapter 3 : Year 1

In order to fully understand what your child is covering in Year One, it is necessary to have a good knowledge of the Reception curriculum so, before reading this chapter, I would strongly advise that the Reception chapter is read. There you will find explanations of fundamental concepts which have not been repeated in subsequent chapters. If your child's birthday is late in the academic year then it is quite possible that they will not be ready to cover certain aspects of the Reception curriculum until Year One when they have matured. Even if your child has covered all aspects of the Reception curriculum, it is likely that they will revisit the work carried out in order to extend and consolidate what they have learnt.

Aspects of the Reception curriculum such as chanting numbers by rote, singing counting songs, counting objects and practical addition and subtraction are still very important. However, it is in Year One that your child will really begin to explore and come to understand the concept of place value (see glossary) and begin to understand some of the relationships between larger numbers. This chapter aims therefore to give you an idea of the types of activity they are likely to be involved in and ideas for ways in which you can help support their learning at home.

<u>Counting</u>

Children will build on the work carried out in the Foundation Stage which has been firmly rooted in practical experience. They will continue to count sets of objects and pictures and will be encouraged to estimate the number before counting to build up their confidence and experience of the size of numbers. The more they estimate, the more accurate their guesses will become. They can be guided by questions such as 'Is it more than 10? Less than 6?'etc. They will initially practise counting sets of up to 20 objects. A good way to help build your child's confidence with larger numbers is to practise rote counting (chanting numbers) both forwards and backwards. This helps children to develop their understanding of the relationship between numbers and is an important skill, needed for solving many addition and subtraction problems. In Year One children will need to practise counting forwards and backwards to and from 20 then beyond. Although they initially find counting backwards difficult, the more they practise, the easier it becomes.

Children will also benefit from being asked to count forwards and backwards from different starting points so don't always count from 0 or 1. You can give them a number from which to count on or back in order to help them become used to counting from different numbers. They can also be asked to stop at a certain number so that they learn to hold a number in their head whilst counting. For example, 'Start counting at 6 and count on until 10/ back until 4'. When they can do this confidently, you can ask them how many they counted (which they can work out by holding up a finger for each number they count).

<u>Ordering Numbers: Numbers to 20</u>

Children are expected to:
- Read, write and place in order numbers, initially to 20 and then beyond, and to know which numbers are greater or smaller than others.
- Identify which number is before or after (or one more/less than) another up to 20 initially, gradually extending this to larger 2 digit numbers as their understanding grows.
- Identify which numbers are in between others, e.g. 'Which number comes in between 6 and 8?' 'Which numbers come between 11 and 15?' etc.

Although there are many different parts of the Year One curriculum, it is fair to say that, at least at the start of the year, much of the number work is based around understanding numbers (and the relationships between them) up to 20. To gain an understanding of these numbers, children need to begin to appreciate the significance of the repeated ones digit in teens numbers. In school it is likely they will be using what are known as place value arrows to construct numbers between 10 and 20. If your child can confidently order and answer questions about numbers up to 10 but is struggling with numbers to 20 it would be a good idea to make some place value arrows of your own (or download some from my website) and to use them to make the numbers between 10 and 20. Using place value arrows allows a child to see that the ones digits are simply having a ten added to them. Once they realise this, many children can simply transfer what they already know about ordering numbers to 10 to the numbers to 20. (Further information about place value arrows can be found in the glossary and on the place value section of this chapter). Numbers to 20 can also be ordered and positioned on number tracks initially and then numberlines in much the same way that numbers to ten are used on number tracks in Reception (see final chapter, 'Games to Play').

Counting in Tens

Children need to begin to appreciate the importance of the 'tens numbers' or multiples of ten in our number system and to begin to recognise the pattern they make (i.e. 10, 20, 30, 40 etc). A hundred square (see glossary) is a good idea to begin with so that the children can actually see the numbers they are counting and the relationships between them. Ten pence coins are also a useful means of working practically with tens. Rote counting forwards and backwards in tens is an important part of developing an understanding of place value. In school children will be given numbers at which to start and stop counting, e.g. 'Count back from 50 to 30. How many did you count?' (Show fingers as count back to represent each ten). This is something you can easily practise at home. It is important as once children can count on and back in tens as well as ones, they will be able to use this skill to help with the addition and subtraction of 2 digit numbers.

As the year progresses and your child becomes more familiar with numberlines, they will begin to see that they do not have to show all the numbers, sometimes only those that are relevant are needed.

E.g. 'Where is 50, 80?' etc

0   100

Figure 3.1

Numberlines can be used to help children to order and position multiples of ten up to 100. Questions such as 'What is the multiple of ten before/after 40?' or 'What is 10 more/less than 60' etc can be asked and multiples can be hidden to encourage your child to work out the multiple between two others and explain their reasoning.

Ordering Numbers: Numbers to 100

Children in Year One will practise rote counting up to 50 then up to 100. Once children are confident with the ordering of numbers up to 20 and the multiples of ten, they will start to do the same sort of activities but with other 2 digit numbers. To gain an understanding of larger numbers, it is likely that hundred squares, numberlines and place value arrows will be used as they were initially with 2 digit numbers up to 20.

Place Value (Tens and Units)

The teaching of place value has changed completely with the introduction of the Numeracy Strategy and is now different in many ways from the work that used to be carried out with tens and units. To start with units are usually now called 'ones' (although some teachers do still use the word 'unit') and no vertical recording will be done at all in the early years. The old method of setting out 'tens and units' vertically and adding the units then the tens should *never* be taught until a range of other strategies have been taught successfully and the children are able to add and subtract 2 digit numbers mentally. [2]

Children are given lots of practical experience of grouping in tens to gain the concept of a 'ten'. They need to be encouraged to count larger groups of objects so that they can begin to see the problems that can occur in terms of accuracy, losing count, time etc. They can then begin to realise the advantages of grouping objects in tens. The old 'sticks of ten' and single units ('ones') may also still be used in schools to represent these. Money (ten pences and one pences) can also be used to help children to understand place value. A small abacus is a good way to show how counting in tens can speed up counting and, because it is already laid out in tens, is a good way to give your child the opportunity to practise counting in tens without having to use a great deal of resources.

'Place value arrows' are a very important resource used in schools for teaching place value. These are small cards with an arrow shape on the right. The arrow shows the tens and the other single digit numbers.

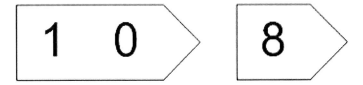

Figure 3.2

These can help children to understand how numbers over ten are made up. In the example above, if the 8 is placed on top of the ten (with the arrows lined up) it will make the number 18. This helps children to realise that the 1 in 18 is actually still a ten (just with the zero hidden). This is very important as some children don't realise the significance of number order so they see 18 as 1 and 8 instead of 10 and 8. This in turn leads to confusions between 18 and 81. These arrows can be used initially to make the numbers 11 to 19 then any 2 digit number. Once children are confident with these smaller numbers, they will start to apply their knowledge to larger numbers, e.g. 36= 30 and 6, 48 is the same as 40 and 8.

It is important that children gain this understanding as it is the basis for much of the later work on addition, subtraction, multiplication and division. If they can split numbers into their tens and ones (partition them) they will find it much easier to work mentally. Place value arrows can be downloaded from www.mathematicsathome.co.uk and I would strongly recommend their use to support your child's understanding.

Once your child is confident with numbers to 20 and multiples of ten, it is a good idea to work with numbers to 30 then 40 etc so that their understanding can be developed gradually. Further work on number to 100 will be continued in Year Two so there is no need to rush your child. It is far better for them to have a secure understanding of numbers to 20 or 30 than a shaky understanding of numbers to 100.

Partitioning

This basically means splitting numbers in ways that make them easier to work with. The most obvious way to partition numbers is into their tens and ones (see above). It is vital that children can do this and understand it, as it is an integral part of many of the strategies taught. As well as partitioning numbers into tens and ones, children are also encouraged to be aware of other ways to split them. They need to develop their understanding of how numbers can be made up through lots of practical work. For example, they should learn that

---

[2] Further explanation of this and the reasons for it can be found on page 73.

7=6+1, 5+2, 4+3 etc through exploration with a range of equipment. As they are working, they need to be encouraged to work systematically and to become aware of the pattern that is being made. For example, when investigating ways to partition the number 5, children may come up with the possibilities in a random order such as 2+3, 0+5, 4+1 etc. It is important to help them to see that if they start with 0+5 and continue with 1+4, a pattern begins to emerge.

0+5

1+4

2+3

3+2

4+1

5+0

Children do tend to spot the vertical pattern 0,1,2,3,4,5 at the start of the number statement and 5,4,3,2,1,0 at the end of the number statement. However, what some of them don't realise is that this occurs because an object is being moved from the right to the left each time and this is what causes the pattern. It is important to help your child to understand this as it deepens their understanding of the process of partitioning and the significance of the pattern. Children who are confident with the ways in which numbers can be split often partition numbers automatically and tend to be better mathematicians because of it. You can help your child to develop an understanding of this by helping them to partition sets of objects practically in the way described above and to help them learn their number facts.

For example, you could use your child's toys to make up a game which involves splitting objects into sets. Together you could collect 5 cars and put them outside a toy garage. You could tell your child that they are going to be a detective who is trying to guess your hidden number sentences. You are going to write all the statements that total 5 and hide them from him/her. Show your child the first statement, 0+5=5 (explain that there are no cars in the garage and 5 outside). See if your child can predict the next 'sum' if a car was moved into the garage then move one car and reveal the sum to see if they were right. You can decide how to score it. You could take it in turns and get a point each for each correct answer or just play it as a guessing game to see how many your child can get right. The idea is that your child begins to realise that as an object is added to one side, one is taken away from the other. If your child finds this hard at first, make it simpler by asking him/her to split the five in different ways in order to find your number sentences. For example, if he/she put 3 cars in the garage and left 2 outside, then reveal the sum 3+2=5 and give your child lots of praise for being clever enough to find it. Now can they find another? Such games should only last five minutes or so and lots of praise should be given. Gradually as they gain confidence, increase the size of the numbers. These sorts of activities build confidence with number and help prepare children for later mental partitioning.

Quick fire testing in short sharp bursts can also be very successful if your child enjoys the challenge of being tested. For example, ask him/her to show you all the ways to make five, using the fingers of two hands. Or ask him/her to tell you all the numbers that add together to make 10. Ask them how they know they have found them all. This can encourage the idea of using the number pattern, beginning with 0+10 and give you the opportunity to show how it can be used (to make sure no number pairs are missed). There are also commercially available games and plenty of activities on the internet which can reinforce these number bonds.

Rapid Recall

Although Year One may seem early to begin learning facts, there are a huge number of facts to learn as the children become older and it is therefore important to learn those that are relevant early on so that they become embedded. They can then be built upon the following year.

Children need to know by heart:

- All the numbers pairs which add together to make 10.

- All the number pairs which add together to make 2, 3, 4 and 5 and work out quickly the corresponding subtraction facts e.g. 2+3=5 so 5-3=2.
- Doubles up to 10+10.
- Halves of any even number up to 20.
- Begin to spot multiples of 2, 5 and 10.
- If confident with the above facts, they can begin to learn all the number pairs which add together to make 6, 7, 8, and 9.
- By the end of the year, they also need to be able to add 3 numbers up to 12 *mentally* using rapid recall of facts to help them.

Although a secure knowledge of the facts listed above will help children to solve problems more quickly and efficiently, it is important to encourage your child to work practically if they need to, particularly if they lack confidence or are encountering a new concept. To use mental strategies, children must have a clear understanding of the concepts involved and practical equipment should always be available to clarify understanding if needed. It is, however, worth pointing out facts such as number bonds, doubles etc while working with practical equipment so that your child will develop a greater awareness of them and eventually be able to spot and use these number facts independently.

Mental Calculation Strategies

These are, as the name implies, strategies which, if employed during calculations, should make them quicker and easier to solve mentally. They often encourage the use of mental recall of number facts and are a vital part of mathematics as they allow children to find ways of working more quickly and efficiently. They also lay important foundations for later work. In Year One the main calculation strategies are:

- If adding, put the larger number first then count on. To use this strategy, children need to be aware that addition can be done in any order (i.e. 3+4=4+3) and they also need to understand the concept of counting on. If they lack this understanding, then they are probably not ready for these mental calculation strategies.

- When adding 3 numbers look for number bonds and add these first (hopefully using instant recall). E.g. 4+2+6, use a number bond to 10: 4+6=10 then add on the 2. This strategy requires rapid recall of number bonds to 10. However, even if your child does not have this rapid recall yet, it is worth pointing out this strategy during homework as it will help your child to see the relevance of number bonds and help them to understand the reason why it is important to learn them.

- Partition (split) numbers to make them easier to work with.

  o Partition 6, 7, 8 and 9 into 5 and a bit. For example, to solve 5+6, encourage children to see 6 as 5+1 then solve as (5+5)+1.
  o Partition other smaller numbers. For example, to solve 8+4, it is useful to know that 4 =2+2 as then if a child knows that 8+2 is a number bond to 10 then they can quickly say 8+2=10. 10+2=12. With smaller numbers, it may not seem to speed up the calculation greatly. However, it is worth bringing the children's attention to this strategy and helping them to understand it, even if they do not use it yet, as it is extremely useful later when working with larger numbers and for later work on partitioning and bridging (see explanation of bridging below).
  o Begin to partition numbers into tens and ones. For example, if adding 14 and 12, it is much easier to partition the 12 into 10+2 then count on a ten and two ones than to try counting on 12 ones. When talking about and introducing the teens numbers, it is a good idea to constantly partition them and recombine them so that children see the pattern that occurs.

- Begin to use 10 (then 20) as a bridge when adding and subtracting, so, for example, when solving 8+5, use a numberline to demonstrate using 10 as a stopping place. Add 2 of the 5 to reach 10 then add the other 3 of the 5 to 10 to reach 13. When initially introduced, this can be shown on a hundred square, as the ten is a natural stopping place. With smaller numbers, this may not seem to speed up the calculation greatly. However, it is worth bringing your child's attention to this strategy as again it is extremely useful later when working with larger numbers.

- Become aware of and recognise doubles and near doubles. E.g 5+5=10 so 5+6=11 and 5+4=9. You can also encourage your child to look for doubles and near doubles when adding three numbers if you feel they are ready. For example, 2+6+6, double 6 is 12 then add on the 2.

- Look for and explore patterns in numbers using practical equipment. Begin to see the relationship between addition and subtraction through this exploration. Pattern in number is very important as it allows children to make vital links during the course of their work. For example, if you know that 5+2=7, then you can begin to work out that 15+2=17 or 50+20=70. It is a good idea to point out patterns that could be useful (e.g. when they begin working with teens numbers) so that they begin to see how patterns in numbers can help them.

- Use compensation. This is when the number is rounded up or down to make it easier to work with then adjusted. E.g. if adding 9 then add 10 and take one off; if adding 11 then add 10 and then another one. This can be demonstrated and practised very effectively using a hundred square as it clearly shows the pattern made by the tens and the effect of adding or subtracting ten.

- Explore the inverse relationship between addition and subtraction through practical work, e.g. 3+2=5 so 5-2=3.

Although at the start of the year, children will still be working with practical equipment and are unlikely to be using mental strategies, it is worth pointing them out in relevant contexts and demonstrating them before your child actually starts to use them so that they begin to gain an understanding of how they can speed up calculations. Later it is hoped that children will begin to use a range of strategies themselves when adding and subtracting.

Addition

It is important to know these basic facts about addition:

- That addition makes numbers bigger (except when adding 0).
- That addition can be done in any order.
- That addition can mean combining two or more sets.
- That addition can mean counting on.

When helping children at home, it is important to talk about adding in terms not only of putting sets together and counting the total (which is the way most often used when working practically with objects) but also as counting on. This can be done through activities such as putting a certain number of objects into a line, for example 5, then adding some more and counting on from 5. For example, 5 socks on a washing line, add 2 more so the total is 5...6, 7. It can also be demonstrated using fingers, e.g. hold up 5 fingers then count 2 more ..6, 7.

The next step is to count on from a number when it can't be seen. So, for example, 5 add 2 could be done with 5 fingers behind the back then count on 2 more or 5 sweets could be hidden in a bag then 2 more added. These types of activities will be understood more readily if they are part of everyday activities. For example, 'There are 5 pennies in my purse. I find 2 more on the window sill, so how many do I have now?' It is important to give these sorts of examples so that your child sees counting on in a context that they understand. A link can be made between these practical examples and numberline work by working practically with a number track with counters (or anything small enough to fit on).

For example, you could say that there were 3 sweets in your pocket and colour the first 3 spaces on the number track (see Figure 3.3). Then ask how many you would have if you had 2 more. You then place 2 more sweets on the number track on numbers 4 and 5 or jump two spaces along with your finger. It is important to discuss the fact that you don't need to count the first 3 as you already know it is 3; you simply count on 2 more from that number.

1	2	3	4	5	6	7	8	9	10

Figure 3.3

Try playing board games such as 'Ludo' or 'Snakes and Ladders' as these are a good way to develop the concept of counting on (and back) on a number track. Whilst playing, attention can be drawn to the fact that you are counting on from the last number and, as your child gains confidence, you can ask them to predict which number (or square if they aren't numbered) they will reach before counting on to find out. Often children find counting on difficult initially so it can be helpful to play large scale games outside with giant dice and numbered stepping stones or number tracks drawn with chalk for them to jump along in order to consolidate their understanding.

It is also important to discuss which number to begin with when counting on. Once your child has a secure understanding that addition can be reversed (i.e. 4+5=5+4), they should discuss whether it is sensible, if given the calculation 8+2 for example, to begin with 2 then count 8 more. This should be explored practically (for example, in the context of a board game that involves throwing 2 dice and adding the total to move on) so that they realise that putting the larger number first is the most efficient way to work when counting on.

When children become confident with counting on using number tracks, a numberline can be used to replace the number track and addition can start to be shown as jumps along a numberline instead.

<u>Numberlines</u>

Numberlines are just a more sophisticated form of number track which consist of a calibrated line. Instead of the numbers being displayed in the actual square, they are written on the calibrations. It is important that before your child begins using numberlines to add (or subtract) they understand addition as counting on (or subtraction as counting back).

The example below shows 8 jumps followed by 3 jumps, which is really addition through combining sets more than through counting on. However, it is useful for some children to see this to begin with so they don't find the work too abstract.
8+3=

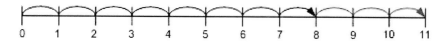

Figure 3.4

As their confidence grows they can begin to count on from 8 as shown below.

Figure 3.5

Children usually begin recording their calculations on pre-prepared numberlines with the numbers already printed upon them. (In Year One, they will also be likely to use a hundred square). In later years, they will record calculations using blank numberlines which they draw.

Once children become confident with the concept of counting on, they can begin to learn how to count on in their head. A way commonly used is 'pocket sums'. The children are told to imagine putting the larger number in their pocket then to count on from that number using their fingers. E.g. 8+3. 'Imagine 8 sweets in your pocket. Now use your fingers to count 3 more...9, 10,11.'

If your child struggles with this concept at first, it can be helpful to put a card in his/her pocket with the numeral on one side and a picture of that number of objects on the other side. He/she could then count on using fingers, cubes, sweets, anything that works! It is important that this work is also shown on a numberline alongside the practical work so that links are made between the abstract and the practical. However, do not overuse this strategy once children are confident with the concept of counting on and number order as some children can become too hooked on counting in ones and it can prevent them from using other strategies.

Although it is not my intention to rewrite the entire Numeracy Strategy for parents, I appreciate that it is useful to have an idea of what your child will be covering in a particular year group and how it should be approached if given as homework. Because of this, I will give a brief summary of the types of calculations that children in Year One will be expected to do in addition and subtraction, along with tips on how to help your child with the examples. It is important to bear in mind that this is only a guide and that, depending on the ability of the child, their age within the year group (maturity rather than ability can sometimes affect the level at which a child is working), their confidence and their previous experience, they may be working at a slightly higher or lower level than that stated. If you find that your child does not understand an example, it is important to give lots of practical work with objects, coins etc and visual supports such as numberlines and hundred squares until they grasp the concept. Then they can move onto more abstract examples. Reducing the size of the numbers in examples is also helpful until the concept is grasped. Sometimes the size of the numbers may be unfamiliar to the children and this affects their confidence and performance.

The following types of calculation for addition will give your child the opportunity to practise their mental strategies, initially with some jottings then eventually using purely mental methods.

1. Add 2 single digit numbers and record the number sentence ('sum')

E.g. 4+5=__   3+__=7   __+2=9   6+7=__

Teach your child to put the larger number first then count on, possibly using a number track, numberline or hundred square to support them if necessary.

Numberlines can be particularly helpful when trying to find missing numbers. For example, 8+__=11 or how many more to reach 11 from 8? Simply begin at 8 on the numberline, then draw the jumps until 11 is reached. How many jumps were drawn?

Figure 3.6

3 jumps were drawn, because 3 more is needed to reach 11 from 8.
If your child cannot grasp the concept of counting on, refer back to examples in the Reception curriculum as it is likely they need to further consolidate their understanding through practical work.

2. Add 3 single digit numbers (within a range of 1-12) and record the number sentence

E.g. 8+2+1=__   4+__+2= 9   __+2+1=6

Remember to look for number bonds to ten (e.g. number 1 above could be worked out 8+2 then add on the 1). This is the most efficient way of working. Doubles and near doubles can also be looked for once your child has some mental recall of doubles facts.

Problems which are open ended rather than lists of 'sums' can be given which will make practising these a bit more challenging for children and help them to develop their confidence. For example, how many totals can you make with these numbers: 3, 5, 6 and 9. Find all the different totals, e.g. 3+5=8, 5+6=11, 9+5=14, 3+5+9=17 etc.

3. Add ten (then 20) to a single digit

E.g. 4+10   3+20

This will test your child's understanding of place value. They may put the larger number first then count on to solve this initially or use ten and one pence coins. Place value arrows may also be used. However, as your child's understanding of place value grows, this should be solved mentally.

4. Add a single digit to ten (then 20)

E.g. 10+5   20+4

Again, this can be illustrated using place value arrows along with practical equipment such as ten and one pence coins. Children may also count on from the 10 or 20 using a hundred square or numberline to begin with. However, once their knowledge of place value is secure, this should be solved mentally without any problem.

5. Add a single digit to a teens number without crossing the twenty

E.g. 14+3

Counting on using 'pocket sums' or a numberline to help would be appropriate. Looking for and using patterns in numbers would also be a suitable way of solving this calculation. If your child knows that 4+3 is 7 then partitioning the 14 into 10 and 4 and discussing the problem would help your child to spot the pattern and use it in other, similar examples. 4+3=7 so 14+3=17.

6. Add single digits to teens (begin to cross the twenty)

E.g. 6+17=

33

Putting the large number first and counting on using a hundred square or numberline could be done. However, once your child has some recall of number bonds, it would also be a good idea to show your child how to use the 20 as a bridge then partition the 6 into 3+3, saying (17+3)+3.

Partitioning the 17 into 10+7 with place value arrows then adding 6+7 (near double) could also be done and is a useful strategy. 10+(7+6) =10+13=23.

This, however, involves a good understanding of place value, partitioning, recombining and near doubles and therefore would only be suggested at this stage if your child were extremely confident with these concepts.

7. Add 2 teens numbers without crossing the tens boundary

E.g. 11+18=

Children are taught to partition (split) the numbers to make them easier to deal with and to encourage (eventually) mental methods. With these smaller numbers, the children can make use of their counting on skills and thus it is not necessary to partition both numbers. It is sufficient at this stage to put the larger number first then partition the second number and count on. In Year One, the children are encouraged to use their own informal methods to record their thought processes and so they should be encouraged to record 18+11 as 18+10+1 on a hundred square or a numberline.

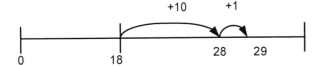

Figure 3.7

They then use their ability to count on in tens and ones to work it out. Although the example above shows a blank numberline, in Year One it is likely that your child will have a pre-prepared numberline that shows all the numbers from 1 to 30 or perhaps 1 to 100. They will then find 18 on the numberline and count on as shown. Place value arrows are very useful for demonstrating how to partition numbers if your child finds this concept difficult. If the arrows are used to make the numbers being added then they can be taken apart to show the tens and the ones so that your child can see how to partition the number to be added.

If you find your child can confidently solve the types of calculations outlined, they can begin to use practical equipment and informal jottings (such as numberlines) to add both single digit numbers and multiples of ten to a wider range of two digit numbers.

Subtraction

It is important to appreciate that subtraction:
- Makes numbers smaller (except when subtracting 0).
- Can be seen as taking away an amount from a set of objects.
- Can be seen as taking away from numbers on a numberline or track (i.e. counting back).
- Can be seen as finding the difference (by counting up).

For example, 6-3
Taking away from a set of objects:

Figure 3.8

or

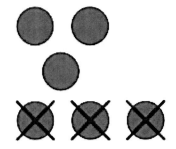

Figure 3.9

Taking away from a number track:

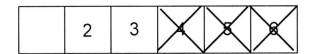

Figure 3.10

Counting back on a number track:

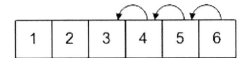

Figure 3.11

Many children find counting back on a numberline far more difficult than taking away from a group of objects. On the number track above (figure 3.10), the numbers are being physically crossed out (i.e. taken away), which corresponds directly with taking away from a set of objects. Children usually understand this quite easily as it links to the subtraction they have experienced when working practically. However, to count back the process is different. It involves jumping back so that the number at the end of the number track does not appear to be counted. For example, to solve 6-3 by counting back (see figure 3.11), a child needs to put their finger on 6 then count back 3 from 6 (i.e. the 6 is not counted). Children usually want to count 3 back including the 6, which leads them to an incorrect answer. If your child is struggling with this then it is best done practically with the children jumping back on large stepping stones or numbers drawn on the ground to begin with. Your child could pretend to be jumping on gold pieces that the pirates are trying to steal. Each time the pirate steals one, it will disappear and so they will have to jump back one to the number before. The idea of jumping back as the one they are standing on is removed sometimes helps them to realise that you count back because the number has been taken away. Once they are confident, they can then move onto jumping back with their finger on a small number track.

Finding the difference:

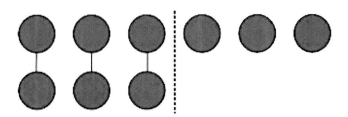

Figure 3.12

It is also important, even at this young age, to understand (in context) subtraction as 'find the difference' and to begin to realise that 'how many more' is the same as subtracting (because it is the same as 'how many less') but may be done in a different way (by counting up instead of back). Using this language and helping children to understand what it means can really help with the later understanding of subtraction when dealing with much larger numbers.

6 is larger than 3 but by how much- what is the difference? To find this out, you need to take away (cover up) the objects/numbers that are the same so that you can count the extra objects/numbers. The difference means the extra objects or numbers. Look at 3 and see how many more there are in 6. This can then be transferred to a numberline and the 3 covered up to show visually the concept of finding the extra numbers which constitute the difference. The question is 'How many jumps is it to get from 6 back to 3?' or 'How many more jumps is it to get from 3 to 6?' It has 3 more so that is the difference. This concept should be introduced through play activities to begin with. A variation on the pairs game can be made which involves selecting two cards (each with a certain number of objects on it) to make a given total. For example, if the object of the game is to make five, then the player that turns over a card with three objects upon it would have to work out how many more are needed to total five and then select the card with that number. Games played with toys (see previous chapter) are also helpful. During games, the process of counting up could be demonstrated on a numberline in order to make the link between the practical and the abstract more obvious. Although this is a difficult concept, once grasped confidently with small numbers, it will help a child's understanding of subtraction immensely.

The following are examples of calculations for subtraction which are appropriate for use in Year One to practise and develop mental methods.

1. Subtract one single digit number from another (without crossing the tens boundary)

E.g. 7-3=

This can be done with practical objects, fingers or by counting back from the larger number, using a numberline to help if need be. 'Pocket sums' can also be used to develop mental methods further.

2. Subtract a single digit from a teens number without crossing the ten

E.g. 16-3.

This can be done by counting back from the large number or by using knowledge of patterns in number: 6-3=3 so 16-3=13. If your child cannot see the pattern, place value arrows again can be used to show that, although a ten is being added, the ones are still the same.

3. Subtract a single digit from 10, then 20

E.g. 10-3=__   20-6=__

Equipment such as money (ten pence coins) can be used to demonstrate this. It is also useful to see the pattern on a hundred square or a numberline.

4.Subtract 10 from a teens number

E.g. 16-10.

Use of practical equipment such as ten pence and one penny coins would be useful here. The use of place value arrows and counting back on a hundred square would also help demonstrate the pattern that occurs.

If you find your child can confidently solve the types of calculations outlined, they can begin to use practical equipment and informal jottings (such as numberlines) to subtract both single digit numbers and multiples of ten from a wider range of two digit numbers.

Number Patterns and Sequences (Multiplication and Division)

Although there is little formal work on multiplication in the form of tables in the Year One curriculum, there is plenty of practical work in the form of solving problems and counting and grouping in twos, fives and tens (see previous chapter for examples). Children will experience multiplication as repeated addition and division as sharing and grouping through practical work, often in a problem solving context.

As well as counting objects in multiples of two, five and ten, they will also rote count forwards in twos, fives and tens and begin to practise counting backwards in these multiples. When counting in twos they will practise counting from different starting points (odd numbers as well as even numbers) so that they see the pattern of the odd and the even numbers. They will continue sequences of numbers and may also record these as jumps on a numberline or by colouring squares on a hundred square. For example, '2, 4, 6.. What are the next 3 numbers?' These types of activities will help children to find patterns in number and are important as they link counting in equal steps to visual images. Looking for and explaining patterns in numbers is very important as it is the basis of multiplication.

Your child will learn to recognise odd and even numbers and begin to recognise 2 digit multiples of 2, 5 and 10. For example, they will start to see that multiples of 5 always end in 5 or 0, multiples of 10 end in 0 and that multiples of 2 end in 2, 4, 6, 8 or 0. You can support your child by chanting in twos, fives and tens with them and looking for opportunities to count objects in twos, fives or tens. Counting 2 pence, 5 pence or 10 pence coins, for example, can help your child to see the need for counting in different multiples. By the end of Year One, children are expected to be able to count on in twos, fives or tens in order to find multiples of these numbers. You can help your child with this by giving real life problems for them to work out, first practically and then by counting on in repeated steps. For example: 'There are six gingerbread men and each one has two buttons. How many buttons would there be altogether?'

Fractions

In Year One children are expected to become familiar with the concept and language of fractions through practical work. In school, they will fold and cut shapes to find halves and quarters, becoming familiar with the idea that to quarter you need to halve and halve again. This idea can be reinforced practically when finding halves and quarters of sets of objects. To divide into 4 equal groups, you can halve the total set, then halve each set again.

By the end of Year One, children are expected to:
- Know that fractions are equal parts of a whole and recognise when a shape is divided into halves and quarters rather than simply cut into two or four pieces.
- Be able to recognise one half and one quarter of shapes and to be able to find a half or quarter of a shape or a small number of objects.
- Be familiar with the notation ½ and ¼.
- Through practical work, they should begin to realise that 2 halves or 4 quarters are equal to one whole. They should also start to realise that 2/4 and ½ are equivalent and that one quarter and three quarters are equal to one whole.
- Gain familiarity with halves on numberlines. For example, recognise that the division between 4 and 5 will be 4½.
- Halve any even number to 20.

<u>Word Problems</u>

These have a very important place within the Numeracy Strategy as they involve putting maths into context. A word problem is basically a real life problem which requires the use of a calculation in order to solve it. Maths can only be of value if it is useful so children need to begin to use their learning in real life contexts and apply what they know to help them solve problems. Children can often find it difficult to decide upon the relevant information in a problem so it is important that children begin to learn to solve problems from an early age. The types of problems that children will encounter are quite simple. However, for some children it is difficult to decide what to do with the numbers – do they add or subtract them? Some children need lots of practice with practical objects in order to gain an understanding of what is required in word problems.

Although problems in Year One will be based initially around addition and subtraction, multiplication and even division problems will also be given which involve combining groups of two, five or ten or sharing or grouping in twos, fives and tens and practical equipment used to solve them. The more experience the children have of the language of word problems and deciding what to do with the information they have, the more confident they will become. Looking at problems together, discussing what they actually mean and then using practical equipment or drawing pictures will help children to decide what they need to do when faced with different questions. They soon begin to realise that the language of word problems will give clues as to what to do. For example, if they are asked 'How many are left?' they come to understand that they need to take away.

In the following problem, for example, fingers, counters or even drawing pictures could be helpful: 'If there are 6 people on the bus and another 5 get on, how many people will be on the bus altogether?' It is not necessary to spend hours drawing detailed pictures; a circle can easily be drawn to represent a person, which is much quicker.

Asking your child hypothetical questions related to everyday activities can help them to gain confidence with word problems. For example, if there are 5 birds sitting on the fence, ask how many would there be if 2 flew away or if 3 more landed on the fence? If your child finds this hard, you can demonstrate counting on or back using your fingers and also draw pictures so that your child begins to make the link between the abstract and the reality of the situation. As children become more confident, it is a good idea to encourage them to make up problems of their own as it consolidates their understanding and builds confidence with the language involved. They can also develop their understanding by making up number stories to go with number sentences (sums) already provided.

In summary, children in Year One build upon previous work from the Foundation Stage. Mental calculation strategies and rapid recall of facts are crucial because these will need to be built upon and added to each year. Talking about their work and explaining their thinking is also vital as it consolidates understanding and allows them to extend their ideas. Practical work is still important but other visual props such as hundred squares and numberlines are also used alongside practical equipment in order to record a mental process. The recording of number statements ('sums') is always done horizontally (see introduction) and children are already being encouraged to use what they already know to work out new facts. In this way, mental methods can begin to develop.

Chapter 4 : Year 2

In Reception and Year One the key skills and strategies of the National Numeracy Strategy will have already been introduced. These skills and strategies are repeated in Year Two at a higher level. Year Two builds upon the work in Year One, so if you find your child has 'gaps' and lacks the understanding to solve some of the Year Two calculations, it is well worth looking closely at ideas from Year One and even Reception, to check that your child is confident with them. In maths it is vital that you start from the level at which your child is working as, if a child has no real understanding of what they are doing, they will be unable to apply and use their knowledge usefully. It may be because a particular concept has been misunderstood that a calculation that relies on this concept cannot be completed. Some children may have a difficulty in a particular area, for example, counting backwards, which may make certain calculations difficult. Improving this one aspect of maths could then result in an improvement in maths generally. In this chapter you will begin to see how the key areas of learning from Reception and Year One are built upon and extended.

Numbers to 100

In Year Two, children increase their confidence with numbers up to 100 and beyond and develop an awareness of where numbers are in relation to each other. This is the year when the understanding of place value (tens and units) is really developed. The groundwork is done in Year One and the children should be confident when splitting numbers below twenty into tens and ones on entering Year Two. They should also have some experience of partitioning 2 digit numbers over 20. However, it is now that the understanding of 2 digit numbers up to 100 is really consolidated.

Children still need to practise estimating and counting larger groups of objects so that they can begin to see the problems that can occur in terms of accuracy, losing count, time etc. This will help them to consolidate their appreciation of the need to group objects in tens.

Counting forwards and backwards in ones, tens and later hundreds (both as a rote count and when counting groups of objects) has a high priority in the National Numeracy Framework and a great deal of time is devoted to it in schools. This is not just to develop an awareness of pattern in number, although it certainly does this. It is actually because it is a skill which is essential for mental addition and subtraction. In Year One, calculations involving addition and subtraction will have often been solved by counting on and back in ones. Once 2 digit numbers are added or taken away, children then need to count on or back in tens and ones. For example, to subtract 12, the most efficient method is to count back ten then count back two. Similarly, when working with 3 digit numbers, counting on and back in hundreds, tens and ones will often be required. Generally speaking, although counting in ones is needed to become confident with number order, it is discouraged when calculating once children know their number facts because use of these facts is quicker. However, until these facts are secure, it remains a useful strategy.

Children need to gain confidence counting up to 100 forwards and backwards and to be able to count on and back from different numbers below 100 (e.g. start from 58 and count back). This can be done orally in short bursts anywhere: in the car, on the way to school. For example, 'Let's see how far we can count before the traffic lights change'. The numbers that cause the most difficulty and therefore need the most practice are those that cross the tens boundary (particularly when counting backwards) e.g. 43, 42, 41, 40, 39. In this sequence, many children would stumble over 39. It is useful for children to practise counting forwards and backwards when they can see the numbers so pointing to numbers on a hundred square as they count can be very helpful. Counting as part of everyday activities can build confidence. As they become more confident, the numbers can be hidden and only used to check if they start to struggle.

An example of a hundred square is shown below. These can be downloaded from www.mathematicsathome.co.uk or they are available to buy as posters from many large stationers and bookshops.

1	2	3	4	5	6	7	8	9	10
11	12	13	14	15	16	17	18	19	20
21	22	23	24	25	26	27	28	29	30
31	32	33	34	35	36	37	38	39	40
41	42	43	44	45	46	47	48	49	50
51	52	53	54	55	56	57	58	59	60
61	62	63	64	65	66	67	68	69	70
71	72	73	74	75	76	77	78	79	80
81	82	83	84	85	86	87	88	89	90
91	92	93	94	95	96	97	98	99	100

Figure 4.1

Hundred squares are extremely useful as they show the patterns made by adding and subtracting ten to numbers if read vertically, as well as showing the pattern of adding or subtracting one if read horizontally.

If your child is struggling counting forwards to 100, then the best way to build their confidence is to work with smaller numbers to begin with then gradually make the numbers larger as their abilities grow. For example, begin by counting to 30 forwards (sometimes using a hundred square to support them, allowing them to see the numbers and patterns) then, once they are confident, begin to go to 40 etc. The same applies to counting backwards. If your child finds this difficult, practise counting back from ten, then twenty and so on, only moving on to a higher number when they are confident with lower numbers.

Partitioning.

As well as recognising 2 digit numbers and being able to chant them, children need to gain an understanding of the relationships between them and the significant patterns that occur. An important part of gaining this understanding is partitioning. Partitioning basically means splitting numbers, often into tens and ones, or hundreds, tens and ones, depending on the size of the numbers. The easiest way to demonstrate how to partition numbers in this way is through the use of place value arrows. They show how numbers can be partitioned into hundreds, tens and ones and can be used to develop understanding, particularly with calculations involving larger numbers. (The teaching of place value and the use of place value arrows is described in the glossary and also in more detail in Year One)[3]. It is important that children can partition numbers confidently because it is the basis for the understanding of many mental and written methods.

---

[3] See page 27.

To partition successfully, your child needs to realise the significance of the position of the digits in 2 digit (and later 3 digit) numbers so check that your child is confident with the number of tens and ones in 2 digit numbers (e.g. knows that 86 is 80 and 6, know that 235 is 200 and 30 and 5). If they have a tendency to muddle numbers such as 18 and 81, for example, then it indicates that they do not fully appreciate the significance of the placing of the tens digit. Using place value arrows to make a range of numbers can help your child to appreciate this concept. Once they can partition numbers in this way, they will be able to work more flexibly (e.g. 50-12 is much easier if seen as 50-10-2). As the year progresses, your child will also begin to gain familiarity with 3 digit numbers and learn to read and write these. Again, place value arrows will be used to demonstrate this. For an explanation of how to use them to develop understanding of 3 digit numbers, see the glossary.

Partitioning refers not only to the splitting of numbers according to the number of tens and ones - it can mean splitting numbers in any way. For example, 7 can be split into 5 and 2. There is usually a great deal of this type of work carried out in Reception and Year One in the form of number bonds. However, if you find that your child lacks the confidence to quickly partition small numbers in their head, then continued practice is required as the ability to partition smaller numbers is needed for using numberlines to add and subtract in the most efficient way. Patterns that occur when numbers are partitioned, such as 0+10=10, 1+9=10, 2+8=10 etc should be investigated and your child encouraged to explain them. So, in the example given, they would try to explain why the first number in the calculation increases by one as the last number decreases by one. They should come to realise that it is because one is moved from the right to the left of the addition each time.[4] Your child could be given general statements to investigate to aid their understanding. For example, 'True or false: 2 odd numbers can never add together to make 10'. The discussion involved in this type of activity can help your child to clarify ideas and develop understanding.

Larger numbers should also be partitioned in a number of ways. For example, children should find ways of partitioning 2 digit numbers. 56 could be seen as 50+6 or 40+16 or 30+26 etc.

Counting in Tens

By now your child should be able to count forwards and backwards in multiples of ten (10, 20, 30 etc). The next stage is to become confident when counting on and back in tens from different starting points, e.g. 11, 21, 31 ...or 15, 25, 35....The more confident they are with this, the easier they are likely to find addition and subtraction of 2 digit numbers.

Again, children will benefit from the support of a hundred square initially to help them see the patterns and relationships between the numbers (by reading it vertically). As their confidence grows over the year, they will be encouraged to visualise it by filling in missing squares when only parts of hundred squares are shown. This helps to develop an understanding of the patterns that occur.
For example,

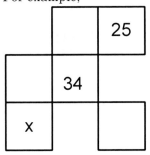

Figure 4.2

---

[4] See page 27 for examples of activities.

A child might be asked to fill in the missing number in the square marked x in the example above. To do this, they would need to be aware of the patterns that exist in the hundred square and know how to use them. Some children may mentally fill in the box below 34 (44) then subtract one, others may fill in the box before 34 (33) then fill in 43 below it. Such methods can lead to discussion which deepens understanding. Eventually, it is hoped that your child will begin to 'see' the hundred square in their mind and visualise the patterns. When this occurs, it should be needed as a prop less and less. Once your child is confident when counting in tens, they should begin to practise counting forwards and backwards in hundreds to begin to develop their confidence with larger numbers. (e.g.100, 200, 300 etc).

<u>Using Numberlines to Order Numbers</u>

Hundred squares and number tracks are some of the first props used to increase your child's awareness of where numbers are in relation to each other. However, numberlines are also of vital importance and become a key strategy for solving calculations and developing mental methods. It is therefore important that your child has a good understanding of numberlines calibrated in different ways so don't always use the hundred square as a prop in calculations. Initially, the numberlines they use will have all the numbers already marked on them, as they generally did in Year One. However, once the numbers begin to stretch beyond 20, it is important that they begin to gain a feel for where numbers are in relation to each other by placing missing numbers on a variety of numberlines calibrated in different ways.

E.g. 'Where is 42, 91?'

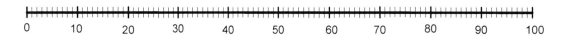

Figure 4.3

If your child is very confident, divisions between the multiples of ten can be removed so that your child can approximate roughly where a given number would go.

E.g. 'Where is 45, 78?'

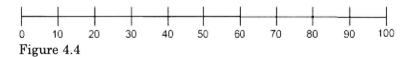

Figure 4.4

Once they can confidently place numbers in this way, they can be asked to find the nearest 10 to given numbers. This is important for later calculations as they will be expected to use the multiples of ten as a 'bridge' so that they can make use of their number bonds.[5] (It is important to remember the rule that numbers ending in 5 are rounded up not down, so 45 to the nearest ten is 50).

To develop competence with numberlines and to become aware of their flexibility, it is important that children see many examples of numberlines with different starting points as one of the advantages that using a numberline has for later calculations is its flexibility. So, for example, when learning about multiples of ten, the numberline below could be used.

E.g. 'Where is 50, 80?'

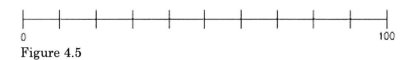

Figure 4.5

---

[5] This is explained in more detail on page 44.

The same numberline could be used when working with multiples of 5 with the starting and finishing number changed. In fact any number could be placed at the beginning and children asked to mark on missing numbers. Of course, they will also need another number marked on (often at the end, although not always) in order for them to work out what the calibrations represent. As they become more familiar with the relationships between numbers, children should be able to approximate on numberlines which have no calibrations. Gaining the confidence to do this will help them understand the use of blank numberlines when they are used later.

E.g. 'Where would 5 go? Why?'

Figure 4.6

<u>Rapid Recall</u>

In Year Two children need to have rapid recall of a number of facts if they are to be able to solve calculations quickly and efficiently. They need to know:

- Number bonds for each number up to and including 10 (addition and subtraction facts) e.g. 1+9=10, 2+8=10, 10-9=1, 1+8=9, 2+7=9 etc. It is important that they are explicitly taught to use these facts to gain other facts. For example, you know 1+9=10 so you also know 9+1=10, 10-1=9 and 10-9=1. You know 2+8=10 so you also know that 20+80=100 and 200+800=1000 etc. You know 10-4=6 so you can work out 20-4=16, 30-4=26 etc.
- Pairs that total 20 (e.g. 0+20=20, 1+19=20 etc). Also, begin to explore and work out quickly other number bonds up to 20 such as 3+15=18.
- Doubles up to 20+20 (then, when secure with these, corresponding halves).
  Again, make sure that you talk about how these can be used to gain others facts. E.g. 3+3=6 so 30+30=60 and 300+300=600.
- Multiples of 10 that total 100, e.g. 70+30 etc. These can be derived from the number bonds to 10. When secure with these, explore the number bonds for other multiples of 10 up to 100, such as 30+40=70, 50+20=70.

<u>Mental Calculation Strategies</u>

Those used in Year One are still used and encouraged but others are also introduced. The strategies encouraged are:

- Consider putting the larger number first when adding two or more numbers by counting on.

- Look out for number bonds to 10 and add these first when adding 2 or more numbers. E.g. 7+3+4 is easier to solve if your child recognises 7+3 as a number bond to 10 as they can then say 10+4.
- Identify doubles and near doubles e.g. if 5+5=10 then 5+6=11 because it is one more and 5+4=9 because it is one less etc. We also know that 50+50=100; 50+40=90; 500+400=900 etc. When your child becomes really confident with 2 digit numbers, other facts can be worked out from these, such as 4+4=8 so 40+40=80 so 40+39=79. Doubles and near doubles can also be looked for and used when adding three or more numbers.

- Recognise patterns in numbers and use them.
  - o E.g. 3+5=8 so 13+5=18 etc.
  - o E.g. 4+3=7, 40+30=70 so 400+300=700.

43

- Partition (split) numbers to make them more manageable or to use them more flexibly.
  - Continue to break 6, 7, 8 or 9 into 5 and a bit. This was recommended in the original Numeracy Strategy as one way to partition numbers in order to work more flexibly. For example, 7+6 could be seen as (5+2) + (5+1). This could then be solved as 5+5+2+1 or 10+3. In the following calculation 8 has been partitioned in this way.
    - 25+8
    - (25+5)+3
    - 30+3=33

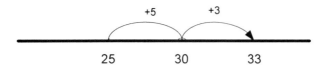

Figure 4.7

  - Continue to partition other small numbers.
    - 27+6
    - 27+(3+3)
    - 30+3=33.

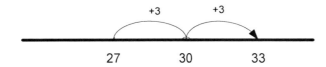

Figure 4.8

  - Partition according to place value. This is when numbers are split into tens and ones.
    - 45+24
    - 45+20+4
    - 65+4=69.

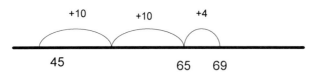

Figure 4.9

- Use compensation. This is when an amount is rounded up or down to make it more manageable, then adjusted at the end of the calculation. For example, when adding 9 (or 11) add ten then take one off (or add one on). This can be done also for 29 (add 30 and take one off) and other numbers near to a multiple of ten. Children in Year Two should practise adding and subtracting near multiples of 10 to any 2 digit number. For example, in the calculation 29+47, the most efficient strategy for children who understand is to take the larger number first then add 29 by adding 30 (count on 3 tens from 47) then subtracting one (because 29 is one less than 30).

- Use multiples of ten as a 'bridge' on a numberline. For example, 82-7 can be solved using 80 as a bridge and partitioning the 7 into 2 and 5.

Figure 4.10

Using multiples of ten as a bridge can be useful whether using large or smaller numbers. For example, 83-25; count back 20 to reach 63 then subtract 3 to reach 60 then subtract the final 2 to reach 58. The 60 is used as a 'bridge' to make it easier to count back from 63 to 58. It avoids the need to count back or forwards in ones and allows children instead to make use of the number facts they already know which eventually encourages the use of mental strategies. Of course, to use this strategy your child must first be able to partition numbers as the usefulness of using the 60 as a bridge relies on knowing that 5 can be partitioned into 3 and 2 and knowing that 63 is 60 and 3 (so therefore 63 subtract 3 is 60). If your child knows their number bonds to 10 then subtracting the final 2 from 60 should require no counting either, as if they know that 10-2=8 then they should be able to work out that 60-2=58. Obviously, there is not only a huge amount of knowledge required to carry out this strategy, but also the ability to apply what they know. A great deal of work on partitioning and using number bonds will be required before they can attempt this type of calculation.

Often children rely completely on their ability to count back in tens and ones. Whilst this is acceptable at the start of the year and is certainly a valid stage in learning how to subtract, some children do tend to over-rely on this, so it worth stressing the use of the bridging strategy (see latter example below) as a quicker method.

For example, 83-25

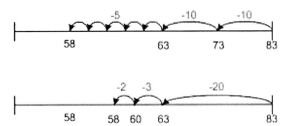

Figure 4.11

- Recognise when two numbers are close together and use this knowledge when subtracting to count up from the smaller number[6] (i.e. to count the difference) instead of subtracting by counting back. For example, 73-68; your child could count up in ones and see that the answer is 5. However, it would be a good idea to use the bridging strategy (using 70 as a bridge so that they can make use of their number bonds and therefore work more quickly and efficiently). 73-68; count on 2 from 68 to reach 70 then count on 3 to reach 73. The total counted on is 5.

Figure 4.12

---

[6] This method of subtracting will be explained more fully later in this chapter in the section devoted to subtraction, page 50.

- To begin to check their work, using the inverse relationship between addition and subtraction, doubling and halving or by adding in a different order or performing an equivalent calculation (e.g. 34-9 could be calculated as 34-4-5 or as 34-10+1).

- Be aware of the inverse operation (i.e. the corresponding subtraction fact to go with an addition fact or division fact to go with a multiplication fact) and be able to use it. This is important as again it will help children to make connections between numbers and work more rapidly in their heads. Thinking logically about the work and using their knowledge about number operations and facts should help children begin to check their work. For example, knowing that addition will result in a larger answer could help a child decide that an error has occurred if an answer was gained that seemed unreasonably small.

It is important to emphasise that we can use what we already know to work out new facts and encourage your child to do this when they can. For example, if your child was counting on from 41 to solve 41+8, then it is appropriate to point out that if they know 1+8 then they won't need to count. Also, make sure that you involve your child in plenty of discussion about their work and ask them to explain how they worked out a certain answer from time to time. This will clarify their thinking and consolidate their understanding of what they have done. It will also give you an insight into the strategies they are using and any misunderstandings they may have.

All of the strategies mentioned will come up and can be used often in later work, right up until Year Five and Six and beyond. It is therefore important that your child gains an understanding of the different strategies early on so that they can be built upon later.

Addition.

Quite often there are a range of appropriate strategies that can be used to solve calculations. Children are encouraged initially to choose their preferred method. However, as they become more capable, an understanding of a wider range of methods becomes essential as children are then in a position to choose the most efficient method which will help them work out answers more easily and quickly.

As in Year One, it is important to know the following facts about addition:

- That addition makes numbers bigger (except when adding 0).
- That addition can be done in any order.
- That addition can mean combining two or more sets.
- That addition can mean counting on.

Before your child can begin to add larger numbers, it is essential that they are able to partition numbers confidently and can recall number facts quickly. Without these skills they will be likely to revert to counting in ones, using their fingers, which is not appropriate for use with larger numbers. It is important that if your child does not seem to understand something initially then relate it to practical examples and allow them to work practically to secure understanding. Also reduce the size of the numbers used, gradually increasing them as your child grasps the concept and begins to understand more abstract examples.

If your child is working confidently with all the aspects of Year Two work, then obviously look at the Year Three chapter to see how to build on what they already know. Beware of simply increasing the size of numbers as a challenge, however. It is far more useful to ask your child to use and apply what they know in the context of a problem if you want to really test their understanding.

The revised framework states that in Year Two, children should work mentally to add and subtract single digit numbers or multiples of ten to or from any 2 digit number.

The following calculations for addition are included in the original Numeracy Framework; it is expected that children should be able to solve these types of calculations mentally using known number facts and their understanding of place value by the end of Year Two. Although these calculations can be solved in a number of ways, I have tried to include explanations and informal written methods that will support the development of mental methods.

1.Consolidate the addition of 2 single digit numbers (crossing the tens boundary)

E.g. 7+5=__

This is included in the Numeracy Framework for Year One. However, it is important to revise this early in Year Two as it allows children to make use of the number bonds they have already learnt. The bridging strategy should be used here; partition the 5 into 3 and 2 then use the 10 as a bridge (thus making use of number bonds to 10). This can be demonstrated initially on a numberline as shown below. As the year progresses, the numberline will become unnecessary for many children and the calculation will be solved mentally.

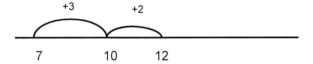

Figure 4.13

2.Add 3 single digit numbers (within the range of 1-20)

E.g. 2+5+6=__

Look for number bonds to 10, doubles or near doubles first and consider whether putting the larger number first would be a sensible choice. In this example, 5+6 is a near double so work this out first then add the 2.

3.Add single digit numbers to a teens number (crossing the twenty boundary)

E.g. 18+5=__

As the 20 boundary is being crossed, use this as a bridge as on the numberline below.

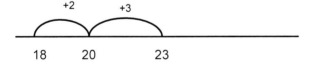

Figure 4.14

Then partition the 5 in order to add it without counting on in ones.

4. Add a pair of multiples of ten (without crossing the hundred boundary)

E.g. 30+30=   ,   30+_=80 etc

Encourage your child to link this back to work on number bonds and doubles if appropriate. We know that double 3 is 6 so double 30 is 60.
From this move onto near doubles: 30+40 is a near double. 3+3=6 so 3+4=7 so 30 and 40 will be 70.

5. Add single digit numbers to multiples of ten

E.g. 3+40

This can be demonstrated by the use of place value arrows, ten pence and one pence coins. Some children may try to solve this by counting on from the larger number ('pocket sums' as explained in Year One may be used). However, if they are still using this strategy in Year Two in this context, it indicates an insecure understanding of place value so plenty of work on place value using the place value arrows is required.

6. Add single digit numbers to multiples of a hundred

E.g. 300+5=305

Again, place value arrows are the most appropriate method to demonstrate the working out of this calculation.

7.Add single digit numbers to 2 digit numbers

E.g. 43+5=___ ,   52+__=56 etc

Your child may solve this by counting on using a numberline as shown below.

Figure 4.15

Start at 43 and draw 5 more jumps then count them 44, 45, 46, 47, 48. This strategy does help them to visualise what is happening and can be useful, certainly at the start of the year. However, it relies on counting in ones which should be discouraged once children know their number bonds, doubles etc. A better strategy to use would be the use of known number facts. We know that 3+5=8 so 43+5=48. It may be useful to use place value arrows to partition the 43 into 40 and 3 if children are finding it hard to make the link with what they already know. They can then see that 3+5 are still being added, regardless of whether the 40 is present or not.

Missing numbers can be worked out using a numberline and this is often the easiest way for children to understand the idea of 'how many more.'

E.g. 52+__=56. Either use a printed numberline to begin with or show your child how to draw a blank numberline, beginning at 52 and ending at 56. Then count on from 52, drawing a jump for each number you count until you reach 56. How many jumps did you draw?

Figure 4.16

4 jumps were needed as you count: 53, 54, 55, 56, so the answer is 4. By the end of Year Two, children should be able to solve this type of calculation mentally. Again, counting in ones is useful to gain an understanding of the concept of the missing number but it is still a good

idea to draw attention to the link to known facts: we know that 4 more would be needed to reach 6 from 2, therefore 4 more would be needed to reach 56 from 52.

It is worth noting that the original framework states that children should add single digits to two digit numbers without crossing the tens boundary. However, the revised strategy does state that single digits should be added to *any* 2 digit number. Obviously, be guided by your child. If you feel they could begin to tackle calculations which do cross the tens boundary, then help them to do so by partitioning numbers and using strategies such as bridging.

8. Add ten (and later multiples of ten) to 2 digit numbers

E.g. 25+10=__
36+20=__

Place value arrows can be used to partition numbers and demonstrate this.
36+20=
30+6+20=
50+6=56.

Alternatively, you could use coins: 10p, 10p, 10p, 1p, 1p, 1p, 1p, 1p, 1p. Put on 2 more 10p. Counting on in tens is an appropriate strategy for a child in Year Two: 36 count on 2 more tens, 46, 56. (A hundred square or numberline could be used to begin with if your child finds this difficult). As your child gains confidence they may also begin to partition the numbers and add mentally. The revised strategy again states that children should add multiples of ten to *any* 2 digit number so, as their understanding of place value and pattern in number grows, children should also begin to solve examples which cross the hundreds boundary.

9.Add 2 digit numbers to a multiple of ten

E.g. 40+36   20+42

In some cases it may be logical to reverse the calculation then count on in tens. However, the use of place value arrows to partition the numbers would be useful here as the children can then add the multiple of ten before recombining the numbers. For example, 40+36 when partitioned becomes 40+30+6. This can then be solved as 70+6. Again, if you feel your child has the understanding then also help them to solve calculations which cross the hundreds boundary.

10. Add a teens number to a 2 digit number (without crossing ten or hundred boundary)

E.g. 43+13

Begin with the larger number then partition the teens number into tens and ones in order to count on.

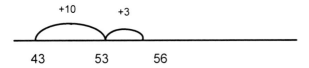

Figure 4.17

11. Add a pair of multiples of 100 (without crossing the thousand boundary)

500+400=__

Make use of known facts and near doubles here. 5+4 is a near double (9) so 500+400 is 900.

49

Subtraction.

As in Year One, it is important to appreciate that subtraction:
- Makes numbers smaller (except when subtracting 0).
- Can mean taking away an amount from a set of objects.
- Can mean taking away from numbers on a numberline or track (i.e. counting back).

It is also important to understand it as 'find the difference' and to know that 'how many more' or 'how many less' is the same as subtraction but may be done in a different way (by counting up instead of back). Using this language and helping children to understand what this means can really help with the later understanding of subtraction when dealing with much larger numbers.

Subtraction is a difficult concept for many children to understand as it can mean counting back or counting up. Initially children will be introduced to it as counting back. The second number will quite often be partitioned and subtracted usually using a numberline as in the example below.

56-12=44

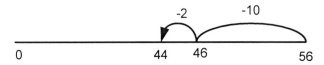

Figure 4.18

This may involve counting back in ones or bridging depending upon the stage at which the children are at. Some children are happy to count back in this way and can easily begin to do this mentally with continued practice.

Subtraction can also be taught as counting up which is the way that many people think of as 'finding the difference'. Although this method seems harder because the number to be subtracted (in the following example, 26) is taken from the start of the number line rather than the end, it is actually easier for many children (and adults!) once they understand it, as it involves counting up or forwards, instead of counting backwards.

For example, 76-26

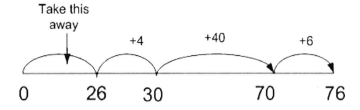

Figure 4.19

Take away 26 from the start of the numberline, so how many are left? Count on to find the answer. It is important to stress that, if taking 26 off 76, you can take it from the beginning (and count up from 26) or the end of the numberline (and count back 26). Although the process is completely different, you are still subtracting the same number from 76 and you will therefore reach the same answer.

Using smaller numbers to begin with can help to get this point across. For example, 5-3 can be seen as:

Figure 4.20

Or

Figure 4.21

You will still reach the same answer wherever the number is taken from. Stress that they could also take the 26 from the start of the numberline then count back from 76 to 26 to find the difference - they would still get the same answer - but it is just easier to count forwards than backwards.

Once children have an understanding of both these methods of subtraction, they can begin to choose which one is more appropriate for various calculations. For example, in the calculation 64-13, it would be sensible to partition the 13 and count back. However, for 64-46, it would make more sense to count up from 46 to 64. It can be useful for children to draw numberlines whilst they are developing their understanding of subtraction as it helps them to visualise the process and decide on appropriate methods.

The original framework gives examples of the following types of calculations for subtraction in Year Two. Children would be expected to use known number facts and place value to mentally solve them by the end of the year.

1.Subtract a single digit number from a teens numbers (crossing ten)

E.g. 16-7=__

Partition the 7 in order to use the 10 as a bridge then count back as shown on the numberline.

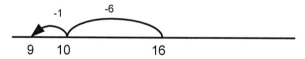

Figure 4.22

2. Subtract a single digit number from a twenties number (crossing twenty boundary)

E.g. 25-7=__

Again, use the bridging strategy; partition the 7 and use the 20 as a bridge.

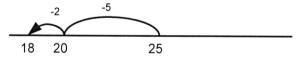

Figure 4.23

Once your child can confidently solve this type of calculation in this way, the numbers can be gradually increased so, for example, single digits could be subtracted from numbers in the thirties or forties. However, it is important that your child is comfortable with the place value

51

of larger numbers and can comfortably work with them without bridging (see example below) before doing this.

3. Subtract single digit numbers from 2 digit numbers

E.g. 46-5=__   79-4=__   52-7=__

Again, your child must be encouraged to use known number facts to work out new ones: we know that 6-5 is 1 so 46-5 must be 41. This could also be illustrated with 10 pence and 1 pence coins or place value arrows at first. They should have a sufficient understanding of place value to realise that 79-4 can be seen solved by partitioning the 79 into 70 and 9 then working with only the ones to say 9-4. This would be practised initially without crossing the tens boundary. Once your child is confident, encourage them to use the bridging strategy (used in the previous two examples with lower two digit numbers) to subtract and cross the tens boundary with any two digit number.

4. Subtract multiples of ten

E.g. 80-40=__

Your child should be encouraged to make links back to known facts again: 8-4=4 so 80-40=40. They may make use of other strategies they have been taught such as counting back in tens, 70, 60, 50, 40. They may even show this on a numberline as below.

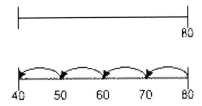

Figure 4.24

Although this can be a successful strategy, it can become time consuming and may be open to errors when counting back larger numbers so it is a good idea as the year goes on to encourage the use of known facts to help with this type of calculation.

The example below also requires subtraction but, as missing numbers in this context are often difficult for children to understand, the numberline is a useful strategy as it can help your child to visualise what is required, thus aiding understanding.

For example, 90-__=50.

A numberline calibrated in tens or a hundred square can used to jump back to 50 from 90.

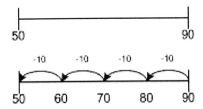

Figure 4.25

How many do we take off from 90 to get back to 50? Count back in tens.
The link to known facts should still be made; so 9-_=5 should also be worked out. As their confidence grows, your child should begin to use these known facts more frequently.

5.Subtract multiples of ten from 2 digit number

E.g. 38-20=

Initially this can be demonstrated using coins: 10p, 10p, 10p, 1p,1p, 1p,1p,1p, 1p, 1p, 1p. Then take off 2 tens. Alternatively, count back in tens 38, 28, 18 using a hundred square or numberline initially as a prop.

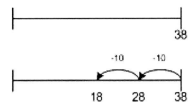

Figure 4.26

Draw 2 tens to subtract then count back in tens from 38. Gradually these props should be removed and children should begin to solve this type of calculation mentally.

6.Find a small difference by counting up (where the 2 numbers lie either side of a multiple of ten)

73-68

Once your child knows where numbers are in relation to each other on a numberline or hundred square, they gain an idea of how close they are and this can be useful for certain calculations. If two numbers are close together it makes more sense when subtracting them to count up from the small number.[7] To understand this, children need to be familiar with the concept of 'the difference'. (Once the concept of the difference is grasped, it can also help with calculations involving missing numbers).

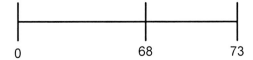

Figure 4.27

Figure 4.27 above shows how either the whole numberline can be drawn or just the relevant part. The same process is carried out whichever way you choose to draw the numberline. The children need to understand that 68 is being taken from 73 but instead of counting back from the end of the numberline, the number is being taken from the start of it. The difference is then counted.

---

[7] See page 50 for further explanation.

Children should be encouraged to make use of the multiple of ten (70) as a bridge and use their knowledge of number bonds to count on. For example,

Figure 4.28

With a little practice using this strategy, they could easily solve this calculation mentally.

7.Subtract a teens number from a 2 digit number (without crossing the tens or hundreds boundary)

E.g. 56-12=

(Take away the ten, then take away the 2).

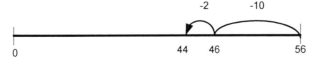

Figure 4.29

This is an example of where counting back is the most efficient strategy. It can be demonstrated initially using a hundred square or a numberline. Your child should be encouraged to partition the 12 into 10 and 2 then count back in tens and ones. It is obviously hoped that they will say 56-10=46. 46-2=44.

8. Subtract a single digit number from a multiple of ten.

E.g. 40-7=__

Working with ten pence coins can help to illustrate the link to the number bonds to 10 and help your child to realise that because 10-7=3 then 40-7=33.

9. Subtract multiples of 100 (within 1000)

E.g. 900-400=__

Link this back to use of known facts: 9-4=5 so 900-400=500 or count back in hundreds.

10.Find what must be added to a 2 digit multiple of ten to make 100

E.g. 50+__=100   30+__=100.

Again, these can be solved by linking the calculations to known facts or using a numberline to count on from the smaller number.

Written Calculations

Although in Year Two the emphasis is on mental calculations, children are encouraged to use practical equipment and informal jottings such as numberlines for calculations that cannot be solved mentally such as the addition of two 2 digit numbers or the subtraction of pairs of 2 digit numbers. For example, 36+22. Children would be expected to partition one number and make use of their ability to count in tens. As addition can be carried out in any order, they would be encouraged to put the larger number first, where appropriate. They would record and work it out on a numberline as shown. In this case the calculation would be 36+20+2.

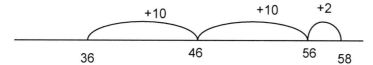

Figure 4.30

Jottings and practical equipment support understanding of the mental process and should be used for any calculation that children cannot solve mentally. In some cases children may be shown how to partition both numbers as an alternative method as follows: 36+22= 30+20+6+2=50+8=58.

Multiplication

As already stated, it is important to recognise doubles and their corresponding halves and to realise that multiplying by two is actually doubling. It is also important to keep pointing out the relationship between doubling and halving.

Children need to be able to count larger groups of objects and learn how to confidently count in 2s, 5s, 10s and 100s (forwards and backwards) so that they can group and count objects in these multiples. They need to estimate amounts before checking by counting in the most appropriate way. The more experience they have of counting and estimating with larger groups of objects, the more accurate their estimates will become so that eventually they will be able to make sensible estimates of up to about 50 objects. You can help your child with this by encouraging them to estimate in context and then check by counting. Depending on the context, it may be appropriate to count in twos, fives or tens when checking. Money is a good opportunity to count in these multiples. Estimates can be made more accurate by asking your child questions to guide their thinking, e.g. 'Do you think there are more than 3 groups of 10 there?'

They will practise spotting patterns between the multiples and become more aware of odds and evens. Again, children need to work flexibly with numbers so once they can count forward in twos and fives they should also practise counting backwards in these steps from 30 or more. It is expected that by the end of the year they will begin to spot multiples of two, five or ten in larger numbers through the realisation that multiples of 2 are even, multiples of 5 end in 5 or 0 and multiples of 10 end in 0. In addition, they should also investigate, continue and explain given number sequences in order to find the patterns that occur. For example, 'Which numbers are missing? 31, 33, 35, 37, __, __.'

To understand multiplication children need to experience it as repeated addition. They must know that, for example, that 4x5 is the same as 5+5+5+5 and be familiar with the language 'sets of' or 'lots of'.

They also need to be familiar with 'arrays' which will help them to see that multiplication, like addition, can be done in any order.

E.g. 3x5

Figure 4.31

is the same as 5x3

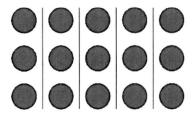

Figure 4.32

As their confidence grows, children should learn how to record arrays and repeated additions as multiplication statements. E.g. know that 5+5+5 can be recorded as 5x3. Arrays can also be used to illustrate the inverse relationship between multiplication and division. So, for example, the array above can be used not only to show 5x3 but also 15 divided by 3. It is important to include plenty of practical work with groups of objects when your child is developing their understanding of multiplication. Once their understanding becomes more secure, they can then begin to work pictorially or use numberlines to support their understanding.

<u>Rapid Recall</u>

Children need to have rapid recall of:

- The facts in their 2, 5 and 10 times tables (and quickly derive the related division facts) E.g. 5x7=35 so 35÷7=5.
- Doubles up to 20+20; derive quickly the corresponding halves.
- Doubles of multiples of 5 up to 50 (e.g. 45+45).
- Halves of multiples of ten up to 100.

They should also be able to record using the x and = symbols and be able to quickly carry out the following:

- Multiply a single digit up to 5 by 2, 3, 4 and 5.
- Multiply a single digit number by 1 or 10.
- Divide a multiple of ten by 1 or 10.

It is important again to encourage children to deepen their understanding by explaining their thought processes. They need to be able to work flexibly and use what they know to work out new (unknown) facts.

For example, if your child is having difficulty working out 9x4 but knows that 10x4=40 then, as long as they are secure with the concept of multiplication as repeated addition, they should quickly begin to understand that if 10 lots of 4 are 40 then 9 lots of 4 will be 4 less than 40. This is much easier than trying to repeatedly count on in fours if they don't know the facts for their 4 times table. Remembering lists of times tables, although important, is not all there is to multiplication. The application of knowledge, particularly in real life problems often requires a good understanding of multiplication. Children need to be able to identify what is needed in the problem, what the relevant information is, and how to solve it in an effective way.

For example, 'Peter has 7 boxes and there are 5 sweets in each box. How many sweets are there altogether?' Although children who know their five times table will be able to solve this mentally as long as they understand what is required to solve the problem, children who don't know the required number fact can be encouraged to work it out by using practical equipment initially to help them. From this, they can move onto drawing a picture to help them, such as seven boxes with a 5 in each, then count in fives to work it out. As they gain confidence they may begin to show their understanding by using a numberline and drawing 7 jumps of 5 then counting in fives to work it out. At this stage the understanding of the problem and the ability to find a meaningful way to solve it are the important factors. Once they have this understanding, similar problems involving larger numbers can be tackled.

Although pictures are often helpful when solving problems, it is wise to encourage your child to use numbers in boxes or jars etc rather than drawing dots (or sweets as in the example above) as children do often become quite hooked on drawing dots and other little pictures. This can become quite a problem as it is so time consuming and inefficient. It also commonly leads to errors, as all the dots that have been drawn have to be counted. It is far better to use a number as, once your child is used to using an actual number to represent an amount, it is only a small step to move onto drawing repeated steps on a numberline which is a far more efficient way to work.

Division

The teaching of division has undergone some major changes since the advent of the Numeracy Strategy. Prior to this, division was often taught mainly as sharing and lots of practical work was done sharing out sweets, biscuits, cubes and anything else available. However, this caused problems for some children with the understanding of division as often division doesn't involve sharing, it involves grouping numbers instead. In fact, the standard methods used to teach division actually involve the repeated subtraction of the same number (or *grouping,* seeing how many groups of that number are in another number). Because the standard methods and many real life problems requiring division involve grouping not sharing, many children found it hard to link the abstract examples to the practical work they had done on sharing. The Numeracy Strategy has addressed this by making sure that teachers teach division as sharing *and grouping.* Children are now given problems such as 30 divided by 5 and taught that it can be solved in two ways. The first can be seen as: '30 (sweets) shared between 5 (people), how many will they get each?' The second can be seen as: '30 (children) put into groups of 5 (for a 5-aside tournament). How many groups/teams will they have?' The answer in both cases is the same. However, the question (and what has to be done practically) is very different.

Once they have been explored using practical equipment such as cubes and counters, division problems can also be solved by drawing simple pictures or illustrated by using a numberline to show how many groups of a number are in another. For example, 'I have 30 sweets and I put 5 in each bag. How many bags will I be able to fill?'

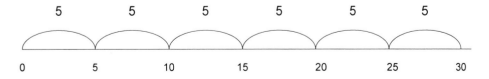

Figure 4.33

Problems requiring division will sometimes have remainders and children need to be familiar with solving calculations which involve remainders. They need to work practically at first until they have a secure understanding of the concept of division, then gain the confidence to use pictorial methods or perhaps numberlines or jottings to work out calculations with remainders.

For example, 32 divided by 5

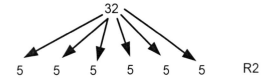

Figure 4.34

You can get 6 lots of 5 from 32, that will equal 30 and there will be 2 left over. It can also be worked out on the numberline as shown below.

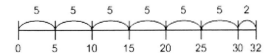

Figure 4.35

Many word problems involve using and understanding division. Division involving remainders does throw up an added complication, as it requires children to be able to decide whether it is sensible to round up or down when remainders are involved.

For example, 32 divided by 5 is 6 remainder 2 but to round up to 7 or down to 6 depends on the problem.

In the first problem it would need to be rounded down:
'I have £32. How many tickets can I buy if they are £5 each?' The answer is 6. There is not enough money to buy 7 tickets.

The second problem needs to be rounded up:
'I have 32 cakes and I want to put them in boxes. I can only get 5 cakes in each box. How many boxes will I need for all the cakes?' The remaining cakes still need a box so this time the answer is 7, although the same calculation was initially required.

The fact that division and multiplication are related (one is the *inverse operation* of the other) is also important and taught early on. The sooner children make this link, the more information they will have at their fingertips. For example if they know that 2x5=10, then they also know 5x2=10, 10 divided by 5 is 2 and 10 divided by 2 is 5. Using numberlines to illustrate division as well as multiplication problems can help children to make this link. However, it is important to ensure that your child has a thorough understanding of the concept through plenty of practical work before attempting more abstract work with numberlines and pictorial representations. Once children have strategies in place for solving problems with smaller numbers, similar problems involving larger numbers can be tackled.

Fractions

Children consolidate their understanding of fractions from Year One and begin to work out halves and quarters of small numbers mentally as well as practically. It is important that they realise that fractions are *equal* parts of a whole. They also need to know that the lower number (denominator) indicates the total number of parts and the top number (numerator) tells us how many parts are represented. They should have a good understanding of equivalences between halves and quarters and know that ¼ and ¾, when added together, are equal to one whole. An understanding of fractions can be gained through practical work involving cutting or sharing objects or diagrams equally. By the end of Year Two, children should be able to find ½, ¼, and ¾ of shapes and sets of objects.

In summary, in Year Two your child will begin to use a wider range of mental calculation strategies for addition and subtraction and increase the number of facts that they can recall by heart through constant use in their calculations. As in previous years, discussion of their work still needs to play a major part in order to help them work through and clarify ideas. They will also gain confidence with the concepts of multiplication and division and learn how to solve problems involving these operations in ways which may well involve using pictorial methods or numberlines. The use of numberlines by now should be firmly embedded so that your child begins to gain the confidence to draw their own empty numberlines to solve a range of problems.

Chapter 5 :   Year 3

In Year Three the calculation strategies employed in Years One and Two are developed further and in many cases employed with larger numbers. The methods of recording these strategies and mental processes do become more sophisticated, although they remain informal and are still, at this stage, very different to the standard written methods taught in schools prior to the introduction of The Numeracy Strategy.

It is often in Year Three that children are introduced for the first time to vertical recording or recording in columns.[8] They should, however, only be introduced to this when they have a firm grasp of the place value of numbers and secure mental skills. It is important that vertical methods are only used when appropriate, that is when the calculation cannot be carried out mentally.

The number of facts requiring rapid recall increases still further in Year Three and it is vital that children are encouraged to memorise them as they can greatly increase the speed at which a child works. This is arguably the area where you as a parent can make the most difference. Learning lists of facts can be rather boring and tedious for children and they need constant repetition and the chance to use them so that they are not forgotten. The number facts learnt in Years One, Two and Three remain important right up until Year Six so, if this is an area with which your child struggles, make sure that they are really confident not only with the facts they are learning in Year Three but also with those required for Year One and Two and revisit these quite regularly so that they retain what they learn.

You can really help by 'testing' your child with quick fire number facts, little and often, to make sure that they retain the information they need. However, it can also be useful to give your child questions to investigate as these can help children to deepen their understanding of the facts they are learning. For example, giving a statement such as: 'True or false: you can add 4 to any number ending in 3 and it will lead to an answer ending in 7,' or 'Is it true that if you add a multiple of 2 to any number it will always give an even number?' Statements such as these require that your child makes up then tests a number of examples, trying to find examples that prove and disprove the statement. It helps them to become familiar with generalisations, explore the patterns in number and encourages them to explain their thought processes. Once they know why a generalisation is or is not true, they can then apply it or make a new rule themselves. When understanding is reached in this way, it is far more likely to be remembered and applied. The facts that need to be memorised for each year group are found in the Rapid Recall section of each chapter.

Place Value

The methods used to help the children to gain an understanding of larger numbers are the same as those used to teach place value (tens and units) in Years One and Two. Place value arrows[9] play a large part as they provide an easy way to demonstrate how 3 digit numbers are constructed. Children can practise making three digit numbers using place value arrows and also learn how to partition numbers into hundreds, tens and ones. This is important because they need to know automatically how many hundreds, tens or ones are in a 3 digit number in order to be able to add, subtract, multiply and divide using mental methods.

To check that your child is confident with the number of hundreds, tens and ones in 3 digit numbers (e.g. knows that 235 is 200 and 30 and 5) ask questions which involve adding and subtracting using their knowledge of place value. For example, 567+30, 421+6, 493-463 all

---

[8]Further details on these methods can be found on pages 5 and 73.

[9] See glossary and page 27.

require an implicit understanding of this concept and, if your child is unable to answer them quickly and easily, they may need to reinforce their understanding. Questions such as 'How many must be added to 452 to make 752?' will give you an idea of whether they have a secure understanding.

When your child implicitly recognises, for example, 452 as 400+50+2 and 752 as 700+50+2, they can then begin to realise that they only need to find the difference between the hundreds (700 and 400) in order to find the difference between both numbers. If your child struggles to answer these types of question, it would be worth checking that their grasp of the place value of 2 digit numbers is secure. Make sure that they do not muddle numbers such as 34 with 43, for example, and that they can add numbers such as 50+3 without counting. If they do need to revise their understanding of place value then work only with 2 digit numbers, initially up to 20 then 30 etc. It is vital to their later learning that they have an implicit understanding that the 6 in 67 is actually a 60 or the 1 in 415 is actually a ten. Once they are fully secure with 2 digit numbers, then return to work with 3 digit numbers. You may be worried that if you spend too much time on this early concept, then your child will fall behind. However, without a secure grasp of the concept of place value and the ability to easily partition numbers, your child will be unable to access many of the mental (and written) calculation strategies that are taught throughout Key Stage Two. It is well worth giving them support at home in this area as it is a fundamental part of later work. Their mathematical ability and understanding will develop far more quickly once they have an understanding of this key concept.

Number to 1000.

In Year Three children are expected to become confident with numbers up to 1000. The emphasis on counting in ones and tens, forwards and backwards, remains from previous years but the starting points get bigger. So, for example, instead of counting on from 52 in ones or tens, children may count on from 152. Children will also be expected to count forwards and backwards in hundreds from different starting points. This ability to count on and back in ones, tens and hundreds is vital for mental addition and subtraction and for much of the informal work carried out in these areas. One successful way to help children gain familiarity with larger numbers and the relationships between them is to practise placing 3 digit numbers on a variety of numberlines and use this as a starting point for identifying one more or one less, ten more/ten less and one hundred more/one hundred less.

E.g. 'Where is 150?'

100   200

Figure 5.1

'Where is 300?'

0   1000

Figure 5.2

As they become more familiar with the relationships between numbers, children should be able to approximate using numberlines without divisions.

E.g. 'Where would 70 go? Why?'

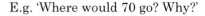

Figure 5.3

'Where is 700? How did you know?'

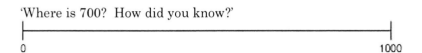

Figure 5.4

Using numberlines (with divisions marked upon them) is also a good way to help your child understand the concept of the nearest ten or nearest hundred. When they are asked to identify the nearest ten or hundred to a given number, they can easily check to see if they are correct. It is important to remember the rule that numbers ending in 5 are rounded up to the next 10 and numbers ending in 50 are rounded up to the next 100. As they gain confidence, they should learn to do this without the prop of a numberline.

Ask your child to state numbers between others on a numberline. Try asking questions such as: 'Which numbers are between 138 and 142?' 'Count back from 156 to 149. How many did you count?' 'Count back 6 tens from 138. How many did you count?' As their confidence increases, in some cases the visual prop of the numberline can be removed and questions answered mentally. However, the use of numberlines and jottings are a useful aid to mental methods and not to be discouraged if your child needs them.

Encourage your child to draw their own empty numberline and to mark a variety of 3 digit numbers upon it. They could begin by marking on multiples of ten or one hundred, then talking about where the other numbers would be in relation to these. However, make sure that your child always has the chance to refer back to a printed numberline if they find this type of activity difficult.

A similar activity involves writing numbers onto cards and asking your child to place them in order. The numbers can initially be consecutive (e.g. put in order: 268, 271, 269, 270) but then become more random (e.g. put in order: 378, 165, 270, 568). Encourage your child to explain their thinking in order to clarify their thought processes and deepen understanding.

Rapid Recall

The ability to remember and make use of number facts during the course of their work is another fundamental aspect of mental calculations. For children to find mental work easy, they need to have a number of facts that they know by heart. Number bonds really speed up mental work and allow children to work much more efficiently. They also give children an alternative to counting in ones, which is time consuming and open to errors.
It is important that by now children have instant recall of:

- Number bonds for numbers up to and including 20 (e.g. 11+9=20, 12+8=20, 20-9=11 etc). They must be explicitly taught to use these facts to gain other facts. For example, you know 11+9=20 so you also know 9+11=20, 20-11=9 and 20-9=11. You know 2+8=10 so you also know that 20+80=100 and 200+800=1000 etc.

63

- Multiples of 5 which total 100, e.g. 45+55, 75+25.
- Number pairs which total 100, e.g. 41+59, 36+64 etc.
- Know multiples of 100 which total 1000 e.g. 100+900, 200+800.
- Know the sums and differences of pairs of multiples of ten (e.g. know instantly that because 9-6=3 then 90-60=30).
- Doubles up to 20+20 (then, when secure with these, corresponding halves) should be known by heart. Again, make sure that these can be used to gain others facts. E.g. 3+3=6 so 30+30=60 and 300+300=600.
- Doubles of multiples of 5 up to double 100 and corresponding halves (e.g. double 35, double 40 etc).
- Near doubles (e.g. if 5+5=10 then 5+6=11 because it is one more and 5+4=9 because it is one less etc. We also know that 50+50=100, 50+40=90 and 500+400=900 etc). Although instant recall of near doubles is not expected, children should be able to spot and work them out very quickly. Later when children are really confident with 2 digit numbers, other facts can be worked out from these, such as 4+4=8 so 40+40=80 so 40+39=79.

The sheer number of facts does seem extremely daunting when first viewed. However, it must be remembered that many of them can be worked out almost immediately from the knowledge they already have. If your child knows their addition facts to 10 and they understand that 20 is actually 2 tens, then learning their addition facts to 20 should actually require very little effort. Similarly, if they know their number bonds to 10 and understand place value then working out 40+60 or 400+600 should again be fairly straightforward. Many of the number facts and mental calculation strategies are linked and it is sometimes the case that because something has not been learnt or understood in one year, it causes problems in later years. All of the above facts will come up and can be used often right up until Year Five and Six and beyond into adult life!

Once your child has rapid recall of a range of facts, encourage them to use number facts during the course of calculations, as, like most skills, the more they are used, the more embedded they become. If your child does not tend to use their knowledge of number facts during the course of their work, demonstrate and regularly point out occasions where they can partition numbers, bridge to ten (see mental calculation strategies, below) or use doubles or number pairs to 10 or 20 in order to raise their awareness of the importance and relevance of these facts. It is important to discuss strategies and encourage your child to make choices about ways to work. It is also useful to you as a parent to ask your child to explain their thinking as it gives you an insight into the strategies being employed during mental work.

Mental Calculation Strategies

As in previous years, there are certain strategies that will allow children to work with increased confidence and flexibility during mental work. Those expected in Year Three are as follows:

- Put the larger number first when adding 2 numbers by counting on, unless another strategy is more appropriate.

- Partition numbers. This is when numbers are split to make them more manageable.
    - 2 and 3 digit numbers can be split according to their place value. For example, double 65, the easiest way is to partition 65 into 60 and 5, then double the 60, double the 5 and add them together.
    - Continue to partition 6, 7, 8, and 9 into 5 and a bit. For example, 45+16. 16 may be easier to work with if it is seen as 15+1 since (45+15) +1 is a relatively easy calculation to solve mentally.
    - It is also vital to learn to partition other small numbers as this is an essential part of the bridging strategy. For example, in the calculation 46+8, 8 may be

easier to work with if it is seen as 4+4 since (46+4) +4 is a relatively easy calculation to solve mentally.

Figure 5.5

If your child finds this difficult, try to encourage them to investigate smaller numbers then help them to use these numbers in calculations which require partitioning. For example, you could ask them to find all the number pairs that total 8.[10] Ask them when they have finished how they can be sure that they have them all and discuss how working systematically, i.e. starting with 0+8 and ending with 8+0, will ensure that they have covered all possibilities. Then give a calculation (such as the example shown on the numberline above) which requires the number 8 to be partitioned. Discuss how because we now know that 8 is equal to 4+4, we can make use of our number bonds to 10 to reach 50 by adding 4 from the 8 to 46 to reach 50. Then we know there will be another 4 of the 8 to add to 50. Make sure that this is demonstrated on a numberline so that your child can visualise the numbers. It is important that plenty of this type of work is done as partitioning smaller numbers is an integral part of mental addition and subtraction work.

- Use the bridging strategy (demonstrated above). This is when multiples of ten and hundred are used as a 'bridge' (or stopping point on a numberline) in order to make use of number bonds and thus calculate more efficiently. The bridging strategy is an integral part of many mental calculations. To use it successfully children need to be able to partition numbers and know their number bonds to 10 as the example below shows.

    27+46. Solve by placing the large number first then partitioning the second number and counting on. Notice on the numberline below (figure 5.6), the 70 is used as a bridge to prevent the need to try and count on in ones from 66 to 73. This bridge is a useful stopping place which aids mental methods by allowing your child to use their number facts to help them and thus solve the calculation more easily.

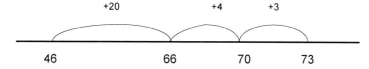

Figure 5.6

- When adding 2 or more numbers, look for number bonds, doubles or near doubles and add these first. (Also, look for pairs that total 9 or 11 in order to add 10 then adjust by 1).

- Doubles and near doubles should be looked for with larger numbers too. For example, if adding 47 and 8, it can be stressed that 7 and 8 are a near double. We know that double 7 is 14 so 7+8 must be one more. 40+15=55. The calculation can now be solved mentally without counting in ones. Similarly, if asked to solve 36+35, it can be pointed out that double 35=70 then add 1. 18+16 can be seen as double 20-2-4.

- Recognise patterns in number and use this knowledge to deduce other facts. E.g. 5+3=8 so 15+3=18, 25+3=28, 55+3=58, 50+30=80 etc.

---

[10] See pages 27 for examples of activities.

65

- Use compensation. This is when an amount is rounded up or down in order to make it easier to work with. Near multiples of ten should be added or subtracted to or from 2 digit numbers.
  - For example, instead of adding or subtracting 99, add/subtract 100 then take one off/ add one back on.
  - 74-39 could be seen as (74 -40) + 1.
  - 45-21 could be solved as (45-20) -1.

  Compensation can also be introduced with 3 digit numbers when adding/subtracting 9 or 11. E.g. 284-9 can be solved as (284-10) +1.

- Recognise that when numbers are close together, finding the difference is the most efficient strategy to use when subtracting. Counting up from the smaller to the larger number is quick and easy to do. E.g. 104-97; instead of trying to take 97 from 104, it is quicker to find the difference between the two numbers. So, use the bridging strategy: add 3 to 97 to reach 100 then add another 4. In total 7 was added so that is the answer. Your child will need to be aware that subtraction can involve counting back or counting on to find the difference. A more detailed explanation of how to approach this can be found in the section explaining subtraction.

- Be aware of the inverse operation (i.e. the corresponding subtraction fact to go with an addition fact or division fact to go with a multiplication fact). This is also important as again it helps children to make connections between numbers and work more rapidly in their heads. If they understand this relationship then they immediately increase the number facts they can work out quickly. For example, they know 14+6=20 so they also know that 20-6=14.

- Using their knowledge about the relationships between numbers can simplify calculations or help children to carry out equivalent calculations. For example, know that the difference between 2 numbers will stay the same if they are both increased or reduced by the same number. 73-48 is the same as 75-50 (as 2 has been added to both numbers) or 70-45 (as 3 has been subtracted from both numbers).

- Children are expected to check their work by performing an equivalent calculation, calculating in a different order or by using the inverse operation.

I have explained the strategies that are used and the facts that are needed in Year Three. The following is an outline of the expectations detailed in The National Numeracy Framework for Year Three for addition. It is worth noting that it is expected that these calculations will be solved mentally using known number facts and understanding of place value by the end of the year. The revised strategy also stresses, in particular, using number facts to 20 to help solve addition involving combinations of single digit and 2 digit numbers. If your child is not yet at the stage of solving these calculations mentally (as may well be the case at the start of the year), then written jottings such as numberlines which support mental methods should be used.

Children are expected to:

1. Add 3 or more small numbers (within the range of 1-50)

E.g. 9+8+14+12=__

Using strategies such as looking for the largest number to add first, looking for pairs that total 10 or 9 or 11 or looking for doubles or near doubles would help to speed up the calculation here.

2. Add 9 or 11 to any 3 digit number then adjust

E.g. 869+9=__ 648+11=__

It is easier to see the 9 or 11 as 10 then add or subtract 1 at the end of the calculations. 869+9 would be seen as (869+10)-1 so that it can be solved more efficiently.

3.Add single digit numbers to 3 digit numbers (without crossing tens boundary)

E.g. 433+5=___   522+__=526 etc

Initially, if your child is still consolidating their understanding of place value, this may be solved by counting on in ones. Children who are used to drawing numberlines may draw the numberline below and count on.

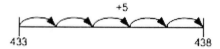

Figure 5.7

However, although this is appropriate at this stage, it is not the most efficient strategy. It is important that children do not become over-reliant on counting in ones so also make explicit the link between 3+5 and 433+5 so your child begins to see the link to what they already know. Once they become very secure with partitioning the number 433 into 400+30+3 they can then mentally isolate the 3 and add 5, thus using what they already know.

522+__=526 can be worked out on a numberline by starting at 522 and finding out how many more jumps are needed to reach 526. For some children, this may involve drawing the jumps from 522 and counting on as you do so. In this case, you would draw a jump from 522 to 523, then another, 524, another 525, and another 526. So in total 4 jumps would be drawn.

Figure 5.8

However, alongside this you need to make explicit the fact that, if you know the difference between 2 and 6, there is no need for any counting on. As the year progresses, it is hoped that your child will begin to replace counting in ones with the use of known facts.

4.Add 2 digit numbers to multiples of hundred

E.g. 34+400

Use place value arrows and partitioning to help your child to develop their understanding of place value so that they learn to solve this type of calculation almost instantly.

5. Add 2 digit numbers to multiple of ten (including crossing hundred boundary)

E.g. 80+34
    60+56

This requires partitioning in order to isolate then add the multiples of ten. This can be done using place value arrows.

80+34 =
80+30+4=
110+4=114

It could also be shown by counting on using a numberline.

Once children are confident with this type of calculation they often partition and use number facts mentally. For example, they may well use number bonds to 100 for the first example and partition 34 into 20 and 10 and 4 mentally, so that they can then see the calculation as (80+20)+14=114. Similarly, with the second calculation 60+56 may be spotted as a near double and solved as (60+50) +6.

6. Add a pair of 2 digit numbers (without crossing tens boundary)

E.g. 36+22=

Your child may put the larger number first then partition and count on the second number. This can be illustrated on a numberline but it is likely that your child will be able to complete the process mentally by now.
36+20+2= 56+2=58.

Another way that your child may solve this is by partitioning both numbers into tens and ones: 30+6+20+2
(Add tens) 30+20=50
(Add ones) 6+2=8
(Recombine numbers) 50+8=58

7.Add ten to any 2 or 3 digit number (including crossing hundred boundary)

E.g. 95+10=__   567+10=__

You can use place value arrows to partition numbers and demonstrate this initially. However, as your child's understanding of place value grows, this should be solved by counting on in tens or using known facts to isolate and add the tens.
 95+10= 90+5+10=100+5=105.
 567+10=500+60+7+10=577.

8. To begin to add mentally 2 multiples of 10 (including crossing hundred boundary)

E.g. 40+70=

Although some children may count on in tens starting with the larger number, it is expected that they will solve this type of calculation mentally and it is therefore advisable to encourage your child to use known facts: 4+7=11 so 4 tens and 7 tens are 11 tens or 40+70=110. Alternatively, encourage your child to make use of their number bonds to 100 by partitioning the 70 into 60 and 10 so the calculation can be solved as (40+60) +10.

9. Find what must be added to a 3 digit multiple of ten to make the next multiple of hundred

E.g. 460+__=500

This can be done by counting on in tens orally or using a numberline initially. However, the knowledge of number bonds to 100 should be made explicit so that once the 60 is spotted (perhaps on a numberline initially), your child will recognise that 40 more must be needed to reach the next multiple of 100.

10. Add multiple of 10 to a 2 digit number (including crossing hundred boundary)

E.g. 95+20   57+70

Again, partition numbers, initially with place value arrows then eventually mentally, to say
95+20= 90+20+5=115
or 57+70=50+70+7=120+7=127.

11. Add multiples of hundred (including crossing thousand boundary)

E.g. 300+300=__   300+__=800
   500+600=__

Counting on in hundreds may be used by some children but using known facts then linking this back to work on number bonds and doubles would be the quickest strategy. We know 300+300=600 because we know that double 3 is 6 so double 3 hundred is 6 hundred. 500+600 is a near double. 5+5=10 so 5+6=11 so we know that 500+600=1100.

12. Add 100 to 3 digit numbers (without crossing thousand boundary)

E.g. 456+100=

This can be done by counting on in hundreds or using place value arrows initially then working mentally to isolate the hundreds and add them.

13. Add a single digit to a 2 digit number (including crossing tens boundary)

E.g. 46+8=

Solve this in 2 steps by partitioning the 8 and using knowledge of number bonds to bridge to the next ten: 46+4+4=54.

Figure 5.9

14. Begin to add two 2 digit numbers (including crossing tens boundary)

E.g. 26+47=
(Partition numbers) 20+6+40+7
(Add tens) 20+40=60
(Add ones) 6+7=13
(Recombine numbers) 60+13 =73

In some of this work number bonds or doubles/near doubles may be spotted which can make the work easier as they can be instantly worked out and then other numbers added to them. This speeds up the addition.

It is worth mentioning that, if your child is finding these Year Three examples difficult, look back to the Year Two examples and check that these can be solved appropriately. Once your child has a secure grasp of the Year Two examples and strategies, they will then have a firm understanding on which to build so that when they return to the Year Three examples, they will be likely to understand them far more easily.

Subtraction.

As in Year Two, children need to understand that subtraction reverses addition (is the inverse operation), that it reduces the size of a number and subtraction of zero leaves a number unchanged.

Subtraction also needs to be understood as taking way from a group, counting back on a numberline and finding the difference between two numbers by counting up (complementary addition).

As discussed in Year Two, subtraction is a difficult concept for many children to understand as it can mean counting back or counting up. Initially children will be introduced to it as counting back, which is necessary for children to gain an understanding of how subtraction makes numbers smaller. It also helps them to see the inverse relationship between addition and subtraction.

To illustrate the inverse relationship 27+20=47 so therefore 47-20=27, for example, the following numberlines can be used.

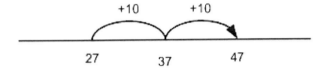

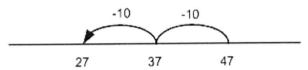

Figure 5.10

However, subtraction can also be taught as counting up, which is the way that many people think of as 'finding the difference'. Although this method seems harder because the number to be subtracted (in the following example, 26) is taken from the start of the number line rather than the end, it is actually easier for many children (and adults!) once they understand it, as it involves counting up or forwards, instead of counting backwards.

So, for example, with the calculation 73-26, instead of counting back as shown below

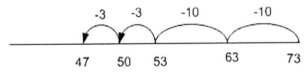

Figure 5.11

You can take the number from the start of the numberline and count up:

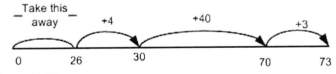

Figure 5.12

70

Take away 26 from the start of the numberline, so how many are left? Count on to find the answer. It is important to stress that, if taking 26 off 73, you can take it from the beginning (and count up) or the end of the numberline (and count back). You will get the same answer whichever method you choose. However, for most people the process of counting up will be easier than counting back.

Your child will need to fully understand counting back before they can gain an understanding of counting on and see the relationship that both have to subtraction. They do need to become confident with both of these methods so that they can decide which is more appropriate for certain calculations.

The National Numeracy Framework for Year Three includes the following types of calculations for subtraction. It is expected that by the end of Year Three children will be able to solve these types of calculations mentally. Again, you will notice that understanding of place value, the ability to partition numbers and use known facts and the use of the bridging strategy are an essential part of many of the calculations. The revised strategy also stresses, in particular, using number facts to 20 to help solve subtraction involving combinations of single digit and 2 digit numbers.

1.Subtract 9 or 11 from any 3 digit number then adjust

E.g. 234-9=__   765-11=__

This involves compensation. Because 10 is easy to subtract, 10 is taken away instead of 9 then 1 added. So 234-9 is seen as (234-10)+1, which is easier to solve mentally.

2.Subtract single digit numbers from 3 digit numbers (without crossing the tens boundary)

E.g. 646-5=__   278-6=__

 Using known facts would be a suitable strategy to use here. We know that 6-5 is 1 so 646-5 must be 641. Place value arrows to partition the numbers before subtracting could be used initially to help children make the link to known facts.

3.Subtract a single digit from a multiples of a hundred

E.g. 400-6=__

Your child needs to be able to partition into hundreds and take away 6 from one of the hundreds then recombine. If they find this difficult, it would be a good idea to practise taking away from one hundred. A pound coin could be used and replaced with 10 ten pence coins to begin with. A hundred square to illustrate the process can also be useful. Children often find taking away from one hundred difficult as they can't visualise it as 10 tens. Practical props such as hundred squares, numberlines and coins can help them to imagine the numbers and thus aid understanding. Also, encourage your child to explain how they worked this out when they complete the calculation in order to clarify and deepen their understanding.

4.Subtract a pair of 2 digit numbers (without crossing ten or hundred boundary)

E.g. 75-42=   98-__=64

This can be done by partitioning the second number and counting back in tens then in ones using a numberline or by counting up, depending upon your child's preferred method and the calculation in question.

5. Subtract ten from a 2 or 3 digit number (including crossing hundred boundary)

E.g. 338-10=   604-__=594

The use of known facts can help with this, e.g. 38-10 is 28 so 338-10=328. When crossing the 100 boundary e.g. 506-10, your child may partition the 10 into 6 and 4 then subtract the 6 to reach 500 then take off 4. This is a valid strategy. However, continue to count forwards and backwards in multiples of ten so that their confidence with the pattern made by counting back ten becomes more evident to them. Practical equipment such as coins (to represent hundreds, tens and ones) should also be used to demonstrate the effect of subtracting ten.

Once they are confident with the pattern made by adding and subtracting ten, missing number calculations such as 604-__=594 should also be easily solved. Coins, place value arrows and numberlines such as the one shown below should all be used to help clarify understanding.

Figure 5.13

6. Subtract a pair of multiples of 10 (including crossing hundred boundary)

E.g. 120-40=

It would be logical to use known facts to work out that because 12-4=8 then 120-40=80. Another way would be to count back in tens from 120. If your child is confident counting back, it would be wise to encourage your child to partition 40 into 20 and 20 then say 120-20=100-20=80. Similarly, your child could count up from 40 to 120, using the 100 as a bridge.

7.Subtract a multiple of ten from 2 digit number (including crossing hundred boundary)

E.g. 68-50=   115-20=

Some children may simply count back in tens from 68 or 115, which is an appropriate strategy. In the above example, it would also be sensible to (mentally) partition the 68 into 60 and 8, then subtract the 50. Once your child is confident working with 2 digit numbers, you could also extend their work into the use of 3 digit numbers. In the latter example, children who are able to work flexibly with numbers may partition the 20 into 15 and 5 then say 115-15=100. 100-5=95.

8.Subtract a pair of multiples of hundred (including crossing thousand boundary)

E.g. 1400-500=

Counting back in hundreds may be used. Known facts can also be used to help: 14-5 =9 so 1400-500=900. However, partitioning (splitting) the 500 into 400 and 100 is probably just as quick for many children. So 1400-400=1000. 1000-100=900.

9.Subtract 100 from any 3 digit number (without crossing thousand)

E.g. 613-100=

Partition the 613 into 600 and 13 then subtract 100 and recombine the number or count back 100.

10.Consolidate subtraction of single digit from teens number (crossing ten)

E.g. 14-8=   16-__=9

Some children would be able to use their knowledge of the relationship between addition and subtraction to count up from the smaller number using 2 steps with ten as the middle stage.

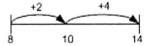

Figure 5.14

For some children, splitting the 8 mentally and taking 4 from 14 then 4 from 10 is just as quick. Either way is an efficient mental method.

11.Subtract a single digit from a 2 digit number (including crossing ten)

E.g. 45-6 =   63-__=58

45-6= can be worked out as 45-5-1.
63-__=58 can be worked out with the knowledge that minus 3 will reach 60 then 2 more will have to be taken off to reach 58. (Counting up from 58 to 63 is equally as valid).

It is important that the tens boundary (or hundred boundary in later work) is used as a 'bridge' or stopping point on the numberline and numbers then partitioned to make them easier to work with. This allows children to work mentally, making use of number bonds which reduces the need for laborious counting in ones.

12. Find a small difference between a pair of numbers lying either side of multiple of 100 (between 100 and 1000) by counting up

E.g.405-397=   602-__=598

Your child needs to realise that when subtracting numbers which are close together it is far easier to count up and find the difference between the two numbers than to count back.

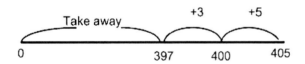

Figure 5.15

13.Begin to subtract two 2 digit numbers (including crossing tens boundary)

E.g. 73-26

This is an example which will probably be solved most easily mentally by counting up.

An Introduction to Written Methods

One of the main reasons for the introduction of the National Numeracy Strategy was to improve mental methods of calculation in schools. Because the aim was (and still is!) to give children secure mental strategies, vertical written methods are not introduced until mental methods are secure. Typically, children should be able to add and subtract a pair of two digit numbers mentally before using vertical written methods of recording. When solving any calculation, they are encouraged always to consider whether it can be solved mentally before attempting a written method. They are also taught to work out an approximate answer before beginning any written calculation so that they will have some idea of the accuracy of their final answer.

The expanded written methods, at first glance, do *look* similar to the traditional methods of working as they are recorded in vertical columns rather than horizontally. However, they are actually very different to traditional methods as children are initially taught to add the most significant (largest) digits first. This is very different from the traditional standard method (which always added the units first) and it often causes a great deal of confusion for parents! The reason for adding larger numbers first is because up until now children have been taught mental methods which involve partitioning numbers and adding the larger numbers first (i.e. tens before the ones etc). The expanded written methods are valuable because they link to these mental methods, encouraging children to work logically and to look at the whole number (and to retain an idea of its size) rather than simply looking at the individual digits. This in turn helps to reduce errors and increases accuracy of work.

For example, in the past children may have solved a calculation such as 203+343 in the traditional vertical format without actually looking at the whole numbers involved. If the actual numbers are looked at a child with a good grasp of place value would be able to solve this very quickly mentally without any sort of written recording. However, many children using the traditional method would fail to look at the numbers involved because they were encouraged to add units, tens etc rather than considering the whole numbers.

Another common error was that children learning how to add vertically would forget to add a digit that had been carried but would be unaware that their answer was wrong as they had no real sense of the size of the numbers with which they were working. For some children with a poor understanding of place value, when a number was carried they would accidentally reverse it so, for example, if they added the units and reached the answer 15, they may actually record the one in the units column and carry the five.
E.g.

```
    3  7
    4  8  +
  _____
  1  2  1
     5
```

Figure 5.16

This would lead to a much larger answer but the child would have no awareness of this. It seems that, if traditional methods are given a very high profile and taught too early, they can actually hinder the development of mental methods and encourage children to simply apply a process to numbers rather than looking at the numbers logically. By teaching expanded methods of recording first, it is hoped that children can gain a full understanding of the numbers with which they are working before moving onto the use of standard compact methods.

Developing Written Methods (Addition)

Written methods really begin informally in earlier years with the use of blank numberlines drawn by the children themselves. They learn how to position numbers on them and how to use them to add and subtract. Their work will have relied upon the ability to partition numbers in a variety of ways, the ability to 'bridge' to the next ten or hundred in order to use their number facts and the ability to use known facts appropriately to gain new facts. However, their recording will have been presented horizontally. The same mental strategies are required for working with larger numbers but, once these skills are secure, children can begin to record vertically using 'expanded' methods of recording. These methods are still used to record or support mental processes. However, many of them also develop understanding of, and eventually lead to, standard written methods. It is therefore important that they are not introduced too soon. They can be introduced from Year Three only if the children using them have secure mental skills (see previous section). Typically, children in Year Three will use written methods to add and subtract combinations of 2 and 3 digit numbers or two 3 digit numbers.

When children begin to record vertically it is essential that they understand the importance of recording hundreds under hundreds, tens under tens, units under units. The significance of this needs to be made explicit with all vertical methods of recording.

In Year Three children may be introduced to expanded methods of addition and subtraction with 2 and 3 digit numbers that cannot be calculated mentally.

Partitioning

Numberlines will have been used as a jotting or aid to support the mental process.
So, for example, 52+37 may have been solved by partitioning the 37 and counting on using a numberline as follows:

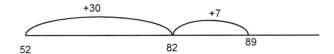

Figure 5.17

Numberlines are very useful for helping children to visualise the process of counting on and back. However, once they are secure with this, the following vertical method may be introduced alongside numberline work in order to familiarise children with vertical recording. It involves the partitioning of only the second number and therefore corresponds directly with counting on using a numberline. The point can be made that either the units or the tens can be added first.

(Approximation: 50+40=90)

```
  5  2            5  2
  3  7  +         3  0  +
 _____          _____
                  8  2
                     7  +
                 _____
                  8  9
```

Figure 5.18

As mental skills progress and as the numbers become larger, it can be useful for children to partition both numbers in a calculation when adding.
For example: 52+37= 50+2+30+7
            = 80+9
            = 89

This method is often quicker and more efficient than using a numberline, particularly when using larger numbers. As children become more adept at mental calculation, they often find they can mentally partition and with smaller numbers may not need to record any of the stages in a calculation like this.

The partitioning of both numbers can be recorded vertically as follows:

```
5  2            50  +  2
3  7  +         30  +  7  +
━━━━━           ━━━━━━━━━━━━━━━
                80  +  9  =  89
```

Figure 5.19

```
367 + 285 =     3 0 0 +   6 0 +   7
                2 0 0 +   8 0 +   5 +
                ━━━━━━━━━━━━━━━━━━━━━━
                5 0 0 + 1 4 0 + 1 2   = 6 5 2
```

Figure 5.20

The examples given in Figures 5.18 and 5.19 are relatively simple and should be solved mentally by children beginning to use vertical methods. Easier examples are demonstrated briefly to ensure understanding. However, these are just an introduction and children very soon move onto more difficult examples involving addition of both 2 and 3 digit numbers.

Both of the methods of partitioning shown above were included in the revised strategy, possibly as a transition stage between the horizontal methods already in use and the vertical layout.

The expanded method below can be slightly more difficult for children to understand initially. However, it is a more efficient method and will be used more as children become more confident with the vertical layout. The example below demonstrates how the hundreds, tens and ones are still added separately but this time beneath each other.

```
3  2
5  9  +
━━━━━
8  0  (add tens)
1  1  (add ones)
━━━━━
9  1
━━━━━
```

Figure 5.21

```
6  7  3
   6  8  +
━━━━━━━━
6  0  0  (add hundreds)
1  3  0  (add tens)
   1  1  (add ones)
━━━━━━━━
7  4  1  (TOTAL)
```

Figure 5.22

As can be seen from the examples, children are taught to add mentally then recombine the numbers to find the total. Instead of adding columns of single digits (as in the traditional standard method), children are actually partitioning the numbers involved then adding them. This reinforces the child's understanding of place value and helps the child gain an idea of the size of the numbers so that, if they gain a ridiculously low or high answer, they will be aware that an error has been made.

It is important to discuss that it doesn't matter which are added first – the tens or the ones. Either way the answer remains the same. Children should develop the confidence and understanding to do it either way. Initially children will be taught to add the largest numbers first as this is what they have been used to doing during mental work. However, as time progresses, children may be encouraged to record the units first in readiness for the standard compact method which they will learn in later years.

In the above example this would mean that the ones would be added (and recorded) first instead of last as follows:

```
6   7   3
    6   8  +
    _____
    1   1  (add ones)
1   3   0  (add tens)
6   0   0  (add hundreds)
_____
7   4   1  (TOTAL)
```

Figure 5.23

Developing Written Methods (Subtraction)

Children will already be familiar with the recording of subtraction on a numberline and it is important that they are confident with subtraction as both counting back and counting up before they begin to record vertically.

Counting Back

Children will already be used to solving subtractions that involve counting back on a numberline.
E.g. 85-63

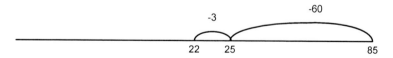

Figure 5.24

The same process can be demonstrated vertically as shown below, in order to familiarise children with vertical methods of recording, once mental skills are secure.

```
    8   5
    6   0   -
  ────────
    2   5

        3   -
  ────────
    2   2
  ────────
```

Figure 5.25

Counting Up (Complementary Addition)

Again, children will know how to count up or find the difference by counting up on a numberline.

E.g. 754-86

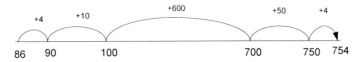

Figure 5.26

 The vertical method below can be introduced alongside a numberline to show how the same calculation can be solved vertically.

```
  7   5   4
      8   6   -
  ────────────
          4   (to reach 90)
      1   0   (to reach 100)
  6   0   0   (to reach 700)
      5   0   (to reach 750)
          4   (to reach 754)
  ────────────
  6   6   8   (Total counted on from 86 to 754)
  ────────────
```

Figure 5.27

Partitioning (Decomposition)

As children gain confidence, they may also be introduced to the expanded method of subtraction by partitioning. It does not correspond directly to any of the mental methods that children have carried out so far, although it does involve partitioning the numbers in the way that children are used to doing. This method does, however, prepare children for the compact standard method of decomposition but introduces it in a way that allows children to retain an understanding of the size of the numbers with which they are working. This means that when they do eventually use the standard method they are far more likely to understand it and make fewer errors.

When working with numbers that require no exchange between tens and units, or hundreds and tens, (i.e. no 'borrowing') this method is very straightforward. However, if 'borrowing' is required then it involves partitioning the numbers in different ways and secure mental skills are essential for it to be carried out comfortably and with understanding.

```
                    300     140   16
   4 5 6           4̶0̶0̶ + 5̶0̶ + 6      3 0 0 + 1 4 0 + 1 6
   1 7 8 -         1 0 0 + 7 0 + 8 -    1 0 0 +   7 0 +   8 -
   ̄ ̄ ̄ ̄ ̄ ̄          ̄ ̄ ̄ ̄ ̄ ̄ ̄ ̄ ̄ ̄ ̄ ̄ ̄    ̄ ̄ ̄ ̄ ̄ ̄ ̄ ̄ ̄ ̄ ̄ ̄ ̄ ̄ ̄ ̄ ̄
                                        2 0 0 +   7 0 +   8  = 2 7 8
```

Figure 5.28

This method requires a great deal of exploration of the ways in which numbers can be partitioned and the fact that hundreds and tens can be moved to allow for subtraction must be discussed. In the example above, 456 is larger than 178 but, when partitioned, there are not enough in the tens or ones from which to subtract without what used to be known as 'borrowing'. This needs to be discussed: there are enough altogether but there are not enough in the tens and ones- what can we do? Hopefully, your child will realise that by partitioning 456 into 300+150+6, there would be enough in the tens but not the ones. Therefore the number needs to be partitioned again into 300+140+16. Many children find this difficult at first and it is worth spending time exploring some of the different ways to partition numbers. So, for example, 456 could be seen as 400+50+6 or 400+40+16. It could also be seen as 300+150+6 or 300+140+16. Moving tens to the units or hundreds to the tens in different ways will help secure your child's understanding of place value and develop their confidence. It is important that they have a solid understanding of this before they attempt to solve subtraction by decomposition.

Numbers which have a zero as a place holder, as in the following example, can confuse some children so make sure numbers such as these are also explored when partitioning.

```
                    600     100
   7 0 4           7̶0̶0̶ +   0 + 4      6 0 0 +   9 0 + 1 4
   5 6 7 -         5 0 0 + 6 0 + 7 -    5 0 0 +   6 0 +   7 -
   ̄ ̄ ̄ ̄ ̄ ̄          ̄ ̄ ̄ ̄ ̄ ̄ ̄ ̄ ̄ ̄ ̄ ̄ ̄    ̄ ̄ ̄ ̄ ̄ ̄ ̄ ̄ ̄ ̄ ̄ ̄ ̄ ̄ ̄ ̄ ̄
                                        1 0 0 +   3 0 +   7  = 1 3 7
```

Figure 5.29

When introducing this method, children should first of all work with numbers that require no borrowing, then move onto examples where numbers from the tens must be exchanged for units, followed by examples where numbers from the hundreds are exchanged for tens. Once children are confident they can move onto examples where hundreds are exchanged for tens *and* tens are exchanged for units. Finally examples which use zero as a place holder should be introduced.

It is worth bearing in mind that vertical written methods are now recommended *from* Year Three. However, because they do rely on secure mental methods, the original strategy stressed the importance of delaying them until mental skills are sound so many schools introduce them later than this. If this is the case with your child's school, don't worry. It simply means when they are introduced, your child will be prepared to use them appropriately.

Multiplication and Number Sequences

In Year Three your child will reinforce their understanding of multiplication by grouping objects practically and looking for patterns in number sequences. They will gain a good understanding of odd and even numbers and test theories about them such as 'All odd numbers end in 1,3,5,7 or 9' or 'If you add 2 odd numbers you will always get an even number'. They will be asked to try to explain general statements and find examples to match them. This exploration of number will hopefully help them to explain and understand patterns in number.

Children need lots of practice counting on and back in steps of 2, 3, 4, 5, 6 and 10. It can be helpful to allow them to colour these number patterns on grids of different sizes, starting from different numbers. This helps them to spot and predict patterns. An example of such a grid is shown (figure 5.30). The grid does not have to a particular size. The example shown is a 5x5 grid. However, it can be interesting for children to colour the same number sequences on a range of grids and discuss what happens to the patterns coloured and why. They will also learn to count on and back in harder steps such as 9 (by adding 10 then subtracting one) and 12 (by adding 10 then a further 2).

It is useful for children to describe number patterns and explain rules, continue sequences (such as 345, 349, 353, __, __) and make sequences of their own. Looking for patterns in number and explaining what they find helps children to think laterally and gives them a deeper understanding of the relationships between numbers.

Understanding general rules about multiples (e.g all multiples of 10 end in 0, all multiples of 5 end in 5 or 0, all multiples of 2 end in 2, 4, 6, 8 or 0, multiples of 100 end in 00 and multiples of 50 end in 00 or 50) will help them when applying their knowledge during problem solving activities. With practice, they should confidently spot 2 and 3 digit multiples of 2, 5 or 10 up to 1000.

1	2	3	4	5
6	7	8	9	10
11	12	13	14	15
16	17	18	19	20
21	22	23	24	25

Figure 5.30

Children need experience of counting on in larger multiples e.g 25s or 50s, and need to be able to answer questions such as 'What is the multiple of 50 before 450?' etc.
As in Year Two, children need to understand multiplication as:

- Repeated addition e.g. 4x5 is the same 5+5+5+5 and be familiar with the language 'sets of' or 'lots of'.
- They also need to be familiar with 'arrays' (see Year 2) and know that multiplication, like addition, can be done in any order so 9x2=2x9; (4x5)x3=4x(5x3) .
- They will start to notice general rules such as multiplying by one leaves a number unchanged and multiplying by zero=0.

80

- It is also important to know that partitioning can be useful to help solve multiplication problems as well as addition and subtraction.
- They must realise that multiplication is the inverse of division (it reverses division) and begin to use this to check their work.

In Year Three children need to gain an understanding of the language and concept of 'scaling'. This is when children have to use multiplication (or division) to work out something that is a number of times as wide, tall, long etc as another. For example, 'Find the string that is 5 times as long as this lace' or 'My bike cost £50 and my brother's cost half as much. How much did his bike cost?' It is important to gain an understanding of this terminology as it is used frequently in word problems in later years.

Children also need to develop their awareness of multiplying and dividing by ten. The grid seen below is often seen in schools as it allows children to see the effect of multiplying and dividing by ten, and clearly shows them the patterns that occur.

1	2	3	4	5	6	7	8	9
10	20	30	40	50	60	70	80	90
100	200	300	400	500	600	700	800	900
1000	2000	3000	4000	5000	6000	7000	8000	9000

Figure 5.31

During Year Three, children will gain confidence with these patterns and learn to multiply single digit and two digit numbers by 10 and by 100. At school, this would probably be demonstrated to begin with using base ten apparatus (blocks of one hundred, sticks of ten and single units). It may also be taught using an abacus (a more traditional method that may have been used in schools well before the Numeracy Strategy) and a calculator (to observe the patterns that occur in numbers multiplied and divided by ten). It can be shown at home in a similar way using money (1 pence, 10 pence, £1 and £10 coins and notes). As long as your child is confident that £1 is equal to 100 pence and £10 is equal to 1000 pence there should not be any particular confusion. A place value board can be made out of a piece of card or paper as shown below:

Thousands	Hundreds	Tens	Ones/units

Figure 5.32

Activities such as asking your child to put 42 pence on the board then multiplying the 40 and the 2 by ten can be done. Each 10p can be multiplied by 10 to equal 100p (£1) so that the 4 tens will be replaced with £4. Then the 2 ones can be multiplied by ten and will be replaced with 2 tens. The calculation is (40x10)+(2x10). If each time the starting number (42) and the answer (420) is recorded then the children should soon spot the pattern. It can be done in the context of money or just discussed as hundreds, tens and units.

Rapid Recall (Multiplication and Division)

Children need to know by heart:
- the facts in their 2, 3, 4, 5, 6 and 10 times tables and should be able to record these using x and =. They should be able to quickly work out the division facts corresponding to these.
- Doubles up to 20+20 and corresponding halves.
- Doubles of all numbers ending in 5 up to 100 e.g. 5, 15, 25, 35...and corresponding halves.
- Doubles of 50, 100, 150, 200, 250 etc up to 500 and corresponding halves.
- Recognise quickly multiples of 2, 5 and 10 up to 1000.
- Recognise 3 digit multiples of 50 and 100.

All of this work will begin with practical work if necessary (although in the case of larger numbers it is advisable to move onto visual representations fairly quickly) then move onto numberlines and finally working out with a few jottings and mentally.

Encourage your child to work flexibly and use what they know to work out new (unknown) facts. For example, once your child appreciates that 4 is double 2 and has a good grasp of the concept of multiplication, they can be taught that they can easily find the facts for the four times table by doubling the two times table. E.g. 2x9=18 so 4x9=36.

This can help when working with larger numbers. To find 4x32, double 32 then double it again. To find 8x32 then double it 3 times (because double 4 is 8). To find quarter of a number the same strategy can be used in reverse - halve the number then halve it again. Similarly, to find out the facts of the six times table, those of the three times table can be doubled.

These little tricks will help your child to understand numbers more fully and work more flexibly with them. However, if they don't grasp new concepts such as this confidently, it is important to illustrate them using smaller numbers and practical equipment. Once your child understands a concept introduced at a simple level, they can then often transfer this understanding to larger numbers.

Mental Calculation Strategies (Multiplication and Division)

Children are expected to:
- Use doubling, e.g. explain that the facts from the 4 times table can be gained by doubling those of the 2 times table. Multiply by 8 by doubling and doubling again.
- Find quarters by halving a half.
- Quickly work out the division facts linked to the 2, 3, 4, 5, 6 and 10 times tables so they can work quickly during mental calculations.
- Use their knowledge of inverse operations to help them work efficiently during calculations and also to check their work (e.g. check addition with subtraction or vice versa, check halving by doubling and check multiplication with division).
- Be aware of the effect of multiplying and dividing by 10. Begin to increase awareness of multiplying and dividing by 100.

The following are examples of calculations for multiplication to be solved mentally (using known number facts and knowledge of place value) in Year Three by the end of the year.

1.Multiply a single digit by 1, 10 or 100

E.g. 3x10   7x100

Children can use known tables facts e.g 7x10=70 so 7x100=700. However, their understanding of place value must be secure in order to be able to apply what they know.

2.Double any multiple of 5 up to 50

E.g. Double 35

Children can partition (double then 30 and double the 5) then recombine the numbers.

3.Multiply a 2 digit multiple of ten up to 50 by 2, 3, 4, 5 or 10

E.g. 40x4   30x3

Knowledge of known facts can be used e.g. 3x3=9 so 30x3=90. Again, understanding of place value must be secure to apply this knowledge. Once your child also knows the facts for the 6 times table, 2 digit multiples of ten could also be multiplied by 6.

4.Multiply a 2 digit number by 2, 3, 4 or 5 without crossing tens boundary

E.g.44x2

If multiplying by 2 the 40 can be doubled and then the 4 can be doubled. If multiplying by 4, you can double the numbers then double them again before recombining them.
The other way to work these examples out would be to partition then multiply as follows:
31x3
(30x3)+(1x3)
 90+3=93
Once the facts for the 6 times table are known, examples such as these which involve multiplying by 6 can be included.

The following are examples of mental calculations for division to be solved mentally by the end of Year Three.

1.Divide a 3 digit multiple of 100 by 10 or 100

 E.g. 300 divided by 10

This also includes finding tenths and hundredths. Although very little will have been formally taught with regard to decimals in Year Three, the work carried out on fractions should allow your child to realise that one hundredth required division by 100 and one tenth requires division by 10. Obviously your child needs to understand place value to solve this type of calculation.

2.Halve any multiple of 10 up to 100

E.g. Halve 40;   Halve 50.

Halving a multiple of ten which has an even number of tens is relatively easy for most children. They simply apply known facts and use their understanding of place value. Half of 4 is 2 so half of 40 is 20. However, sometimes children struggle to halve a multiple of ten if it has an odd number of tens, such as 30, 50 etc. If your child struggles with this, they need to be shown that 50, for example, is made up of 5 tens. Then they can halve the 4 tens and discuss how to halve the last ten. Practical equipment, such as ten pence and one penny coins can be useful to demonstrate this concept.

Developing Written Methods (Multiplication)

In Year Three, children will begin to multiply 2 digit numbers by single digit numbers. They often begin by using informal methods such as drawing pictures or numberlines then using repeated addition to add the tens then the ones. Once the concept of multiplication has been

consolidated with larger numbers, they can then use informal methods to partition then multiply the tens then the ones.
For example, 32x3;

This may initially be solved by using a numberline to show three jumps of 32.

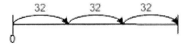

Figure 5.33

Then repeated addition could be used to work out 30+30+30=90 and 2+2+2=6. So the answer is 96. Some children may initially feel happier drawing pictures to help them make the link between practical contexts and the abstract. So, for example, if they have been set the problem: 'There are 3 boxes and each contains 32 sweets. How many sweets are there altogether?' they may actually want to draw 3 boxes. It is acceptable to draw 3 boxes and write the number 32 in each.

Figure 5.34

This is still an efficient way of working. The numbers can still be partitioned and the calculation solved by repeated addition as on the numberline. However, actually drawing 32 sweets in each box (as some children try to do) is not acceptable; it is time consuming, open to errors and does not allow children to make use of any knowledge or skills they already possess (other than drawing and counting!)

From repeated addition, either in the form of a numberline or pictorial jottings such as boxes, children can easily make the transition to 30x3=90 and 2x3=6, as shown below.
32x3=
(30x3) + (2x3)
90+6=96.

The children may record it in a way similar to the above. However, the informal 'grid' method (introduced in the original Numeracy Framework in Year Four) is now suggested for use in Year Three too. This involves partitioning 2 digit numbers and multiplying each part as shown below.
E.g. 32x3=90+6=96

x	30	2
3	90	6

Figure 5.35

These informal 'jottings' can help children to show and understand their thought processes. They are also encouraged to verbally explain their work in order to clarify their thoughts and methods.

You will have noticed that, as with addition and subtraction, in multiplication partitioning the numbers is fundamental to the development of mental methods. It also encourages the children to gain an idea of the size of the numbers they are working with so that they are more likely to recognise an answer that is far too high or too low.

Developing Written Methods (Division)

As already explained in Year Two, the teaching of division has undergone some major changes since the advent of the Numeracy Strategy. Prior to this, division was often thought of primarily as sharing and lots of practical work was done sharing out sweets, biscuits, cubes and anything else available. However, this caused problems for some children with the understanding of division as often division doesn't involve sharing, it involves grouping numbers instead. In fact, the standard methods used to teach division actually involves the repeated subtraction of the same number (or *grouping,* seeing how many groups of that number are in another number). Because the standard methods and many real life problems requiring division involve grouping not sharing, many children found it hard to link the abstract examples to the practical work they had done on sharing. The Numeracy Strategy has tried to address this by advising that teachers teach division as both sharing *and* grouping much earlier. Children are now given problems such as 10 divided by 5 and taught that it can be solved in two ways. The first can be seen as '10 (sweets) shared between 5 (people), how many will they get each?' The second can be seen as '10 (children) put into groups of 5 (for a 5-aside match). How many groups/teams will they have?' The answer in both cases is the same. However, the question (and what has to be done practically) is very different.

Many word problems involve using and understanding division. Division involving remainders does throw up an added complication, as it requires children to be able to decide whether it is sensible to round up or down when remainders are involved.

For example, 32 divided by 5 is 6 remainder 2 but to round up to 7 or down to 6 depends on the problem.
 In the first problem it would need to be rounded down:
'I have £32. How many tickets can I buy if they are £5 each?' The answer is 6. There is not enough money to buy 7 tickets.

The second problem needs to be rounded up:
'I have 32 cakes and I want to put them in boxes. I can only get 5 cakes in each box. How many boxes will I need for all the cakes?' The remaining cakes still need a box so this time the answer is 7, although the same calculation was initially required.

The fact that division and multiplication are related (one is the *inverse operation* of the other) is also important and taught early on. The sooner children make this link, the more information they have at their fingertips. Using numberlines can help children to visualise this link. For example, given the problem: '24 cakes are put in boxes. 4 cakes will fit in each box. How many boxes will be filled?' the following numberline can be drawn to work out the answer.

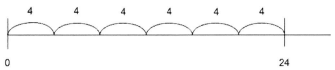

Figure 5.36

As the jumps of 4 are drawn, your child can count in multiples of 4 until they reach 24. They then count the jumps (which represent the boxes) to find out how many there are. This method clearly shows how division is linked to multiplication. Again, if this is too abstract to

85

start with, your child can be encouraged to draw their own representations (perhaps boxes with a 4 in each instead) to represent the problem.

Once they have an understanding of division, children who know their tables will easily solve division problems. For example, if your child knows that 2x5=10, then they also know 5x2=10 and 10 divided by 5 is 2 and 10 divided by 2 is 5. In Year Three your child will begin to develop the understanding that, unlike multiplication, division cannot be done either way (i.e. 10 divided by 2 is not the same as 2 divided by 10).

<u>Fractions.</u>

Fractions are introduced and taught in a practical context to begin with and initially children will carry out work involving cutting, folding or shading parts of shapes and dividing small groups of objects. From this, they can move on to ordering fractions in practical contexts so that they gain a sense of the size of fractions in relation to each other. Children will be expected to:

- Understand that the denominator indicates the number of parts into which the whole has been divided and the numerator indicates the number of parts in question.
- Identify and estimate fractions of shapes.
- Read and write proper fractions (e.g. 2/3, 4/5) and find unit fractions (e.g. ¼, 1/3 etc) of quantities. For example, they should be able to find half of a quantity by dividing by 2, or a third by dividing by 3.
- They should be able to compare fractions on diagrams and begin to recognise equivalences. For example, by looking at a diagram see that 1/2 is equal to 2/4 and 3/6 etc.

It is important that if your child does not seem to understand something then, as in previous years, relate it to practical examples and allow them to use equipment to help them. It is also helpful to reduce the size of the numbers used, gradually increasing them as your child grasps the concept and begins to understand more abstract examples.

In summary, in Year Three children continue to build on the mental strategies introduced in Year Two but also develop a wider range of recording methods. New number facts are learnt and those from Years One and Two are consolidated and practised repeatedly during the course of calculations. At the start of the year, children will continue to consolidate their understanding of number relationships and mental strategies by drawing their own blank numberlines to solve calculations. As the year progresses (depending upon their understanding), they may also be introduced to vertical methods of recording addition and subtraction calculations for the first time. These vertical methods teach the children to use many of the same strategies but to record their calculations in a new way, in preparation for the standard written methods in later years. They will be encouraged to use what they know to make approximations of their answer before starting work and to check their work on completion. Multiplication is explored in greater depth with larger numbers and non-standard methods of recording are introduced. The concept of division is consolidated and the relationship between multiplication and division is explored, as are the relationships between various times table facts. Although the examples given tend to concentrate on how to solve given calculations, it is worth remembering that children must be able to apply their skills in problem solving situations. Make a point, therefore, of asking your child to use the skills they have to solve problems, or setting calculations in a problem solving context, rather than giving lists of calculations to solve.

Chapter 6 : Year 4

By Year Four, the key mental strategies have been introduced and children now begin to use them, not only with larger numbers but also with negative numbers and decimal numbers. Because of this, it is not always easy to solve calculations mentally and written methods of calculation are therefore required on a more regular basis. The emphasis on mental methods is still present, however, and children are always asked to consider whether a calculation can be solved mentally before employing written methods. In this chapter I will continue to explain the mental strategies and the contexts in which they can be employed but I will also explain the written methods that are now used and try to give an understanding of the advantages that they have over the traditional written methods.

Place Value

At the start of Year Four it is expected that children will have a secure understanding of place value (thousands, hundreds, tens and units) with 3 digit numbers and will be able to partition (split)[11] them and recombine them confidently. This understanding needs to be secure as it is fundamental to both the development of mental methods and the ability to use the informal written methods for addition, subtraction, multiplication and division in Years Four, Five and Six. Assuming that these skills are in place, in Year Four the relationships between 4 digit numbers are explored more fully and the understanding of place value is extended to include decimals. Place value arrows[12] continue to be used to demonstrate the relative value of digits in positive whole numbers and can show how four digit numbers can be partitioned and recombined. Children need to understand, for example, that 3428 can be partitioned into 3000+400+20+8 if they are to be able to work with that number. Children also need to order, read and write 4 digit numbers in order to develop their understanding of them. You can help your child by asking them to place 4 digit numbers on numberlines.

E.g. 'Which number is half way between these two numbers?'

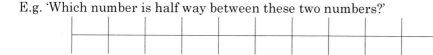

4000   4100

Figure 6.1

You can also ask them to state numbers before, after or between others and to round 2 and 3 digit numbers to the nearest ten or hundred (knowing that they round up if they end in a 5 or 50). If they find these questions difficult to answer, allow them to find the numbers on a numberline. Once they gain confidence, the numberline can be removed.

Once they can place 4 digit numbers confidently on numberlines, they should be encouraged to estimate numbers on uncalibrated numberlines. E.g. 'What number do you think this is? Explain how you know?'

0   100

Figure 6.2

---

[11] See glossary for further explanation.
[12] See glossary.

87

By changing the numbers at the start and the end of a numberline, the same question can be asked to develop an understanding of the relationships between a wider range of numbers. E.g. 'What number do you think this is? Why?'

Figure 6.3

Counting forwards and backwards from different starting points is as important as in earlier years. Filling in spaces on numberlines can help consolidate oral counting on and back and help build confidence with ordering 4 digit numbers. For example, children could be asked to count on or back from 3428 or they could be asked to fill in the spaces below.

				3428					

Figure 6.4

Children need to be able to count forwards and backwards from 4 digit numbers not only in ones but also in tens, hundreds and thousands. This will further develop their understanding of place value and pattern with larger numbers. For example, to count on in tens from 3428, they need to realise that the 3000, 400 and 8 will initially remain unchanged and only the tens will be affected until the hundreds boundary is eventually crossed.... 3428, 3438, 3448, 3458, 3468, etc. If your child has carried out and understood the place value work based around 2 and 3 digit numbers in Years Two and Three, then this work is a natural extension and should pose no particular problem, as long as it is introduced and explained appropriately and children are given the chance to practise the skills they have with these larger numbers.

You can find out whether your child is developing a secure grasp of the place value of 4 digit numbers by asking them to solve examples such as the following:
2004-7= __
4576+200=__
7692-500=__
3428+____=8428
8513-__=8213
If your child finds the pattern of the numbers difficult to grasp then, as well as using place value arrows to consolidate their understanding, it may also help to use some sort of practical equipment. In schools it may well be the traditional place value equipment. However, the same purpose can be served at home using money. One penny coins obviously represent ones, ten pence coins represent tens, one pound coins represent hundreds and ten pound notes represent thousands. As long as your child is made aware that the pounds actually represent one hundred pence and the ten pound notes are really one thousand pence they should be able to appreciate the concept and add one pence coins, ten pence coins, one pound coins and ten pound notes. Plastic money sets can be used or pretend £10 notes can be made if you don't happen to have piles of ten pound notes around the house!!

It is wise to begin with smaller numbers (e.g. show the patterns of adding tens and hundreds in two and three digit numbers first) before trying to explain patterns in four digit numbers. If your child is struggling when working with smaller numbers then it is important to keep practising counting on and back and adding/subtracting these numbers until they are confident. Try to resist the temptation to move onto larger numbers if your child is not really ready, as this may undermine their confidence and thus delay their understanding. Once an understanding is reached with smaller numbers, this can often be applied quite quickly to larger numbers so time spent developing an understanding of the basics is time well spent.

Digit cards are also useful to consolidate the understanding of place value. These are cards with single digit numbers on (from 0-9). Tasks such as 'Make the largest 4 digit number you can with these cards' will help children to realise that the thousands is the digit that has to have the highest number in it. Other instructions (such as: 'Make me a number between 5000 and 6000'. 'Make me a number greater than 2500'. 'Make me a number that contains 5 hundreds.') can be used to check your child's ability and understanding. It is important to make sure your child can work confidently and flexibly with 4 digit numbers and truly understands what each digit represents so that they are able to partition them mentally in order to solve problems. Discussion and explanations can also play a large part in developing understanding and confidence during these type of activities. For example, 'How can you be sure that this is the largest number you can make?' can help your child to clarify their thought processes and give you an indication of their understanding.

Negative Numbers

Just as with positive numbers, you can help your child to gain an understanding of negative numbers through counting, ordering them and placing them on numberlines. They are often introduced in the context of the weather (practical work with strip thermometers is sometimes carried out to help children gain confidence with measuring temperature). They are also often displayed upon vertical numberlines in pictorial form with the positive numbers above ground, zero being ground level and negative numbers below the ground. Sometimes the context of a frozen pond is used, with negative numbers shown below the surface of the pond. (It is worth remembering that numberlines are not always displayed horizontally. They may be shown vertically too and it is important that children appreciate this).

You can help your child to count backwards to include negative numbers (e.g. 3, 2, 1, 0, negative 1, negative 2 etc) and to count on from negative numbers (e.g. -3, -2, -1, 0, 1, 2 etc). Ask them to use numberlines to find integers (whole numbers) between 2 others. For example, 'Which numbers are between -5 and 4?' They need to gain confidence ordering numbers that include both positive and negative numbers both from smallest to largest and from largest to smallest. Writing a range of integers on cards and helping your child to order them should help them to realise that negative numbers work in reverse to positive numbers, i.e. the larger the numeral, the smaller the number. For example, 5 is smaller than 10 but -5 is larger than -10. For some children this can be initially rather confusing which is why visual prompts such as numberlines and thermometers are so useful: they allow the children to see where the numbers are in relation to each other so that children can begin to visualise them.

Children in Year Four will be expected to state which is the larger or smaller of two integers and to use the greater and less than signs to compare negative and positive numbers. For example, they may be asked to fill in the missing answer in the following: ___> -5. As their confidence increases, they should also begin to predict the next number in sequences that extend beyond zero and recognise negative numbers on calculator displays. They should develop their understanding through discussion of problems involving negative numbers. For example, 'The temperature is minus 2 degrees celcius. By how much must it rise to reach 3 degrees celcius?'

Decimals

Decimals are taught in a very similar way to whole numbers. Children need to gain an idea of the size of decimal numbers and the relationship between them. To do this, they need to realise that the decimal point separates the whole numbers from parts of the whole and then extend their understanding of place value in order to recognise what each digit after the decimal point represents. To begin with, try to link decimals to the children's own experiences so that the work is not too abstract. Money and measurements are good contexts for teaching about decimals as they are contexts with which most children are familiar. It is relatively easy to understand and work out that 10 pence is a tenth because it is a tenth of a pound. Similarly, one penny is a hundredth of a pound. This understanding can be developed further by partitioning and recombining decimal fractions, initially in the context of money or measurements, then, as confidence grows, in more abstract contexts, e.g. 4.51 = 4.00+0.5+0.01. Reverse place value arrows (showing units, tenths and hundredths) can be used to help children to partition and recombine decimals numbers. For example, on the arrows shown below, the 0.3 would be placed on top of the 0.01 so that the 3 covers the 0 in the tenths column, then the 9 would be placed on top of them both covering the zero in the units column, thus making the number 9.31. (The arrows on the left should line up on top of each other to indicate that the digits are correctly placed).

Figure 6.5

Just as with whole numbers (integers), children need to practise counting forwards and backwards to develop an idea of number order so ask your child to count up in tenths from zero then back again. Initially use the context of money; for example, ask your child to count forwards and backwards in steps of ten pence or one pence from different starting points. Again, to begin with, supply a numberline to count along and remove it as your child becomes increasingly competent. Always allow your child to check using a numberline if they feel the need as removing aids such as this before a child is ready can undermine confidence and hinder understanding. Once your child's confidence grows, they can be asked to count on and back in decimal steps of different sizes (e.g. steps of 20 pence), to predict the next number in a sequence of decimal numbers (such as amounts of money) and to describe and explain patterns in number sequences involving decimals.

To develop confidence with decimal numbers, children also need to become used to ordering them and you can help with this by writing a range of decimal fractions on cards (to begin with use the context of money or measures until they gain confidence) and asking your child to place them in order from smallest to largest or largest to smallest. Discuss how they know which is the smallest/largest, using reverse place value arrows to check and clarify their understanding.

Problems often require children to be able to convert between units, such as converting pence to pounds or centimetres to millimetres so it is important for your child to practise ordering mixed units when working with money or measurements. For example, they may be asked to order: 50cm, 2.5m, 520cm and 5mm. They should discuss what these amounts mean and how they can be converted to a common notation using a decimal point.

Ask your child to place decimal fractions on numberlines calibrated in different ways. Again, to begin with use the context of money or measurements. As they gain confidence, the work can become more abstract. For example, show the numberline below and ask 'Where is 6.1?' Encourage them to explain how they worked out the correct position and link back to the context of money or measurement if your child has difficulty.

Figure 6.6

Using the numberline as a visual prop, they can be asked to identify decimal fractions before, after and in between others. For example, 'A CD costs between £5.80 and £6.40. What could it cost?' 'Which decimal fractions lie between 5.8 and 6.4?' or 'What is the decimal fraction before 6.12?' They should also learn to round decimals with one and two decimal place to the nearest whole number (this can also be done by looking at the number on a numberline to start with) and money should be rounded to the nearest pound. Once your child becomes confident when answering these types of questions using a numberline, this visual prop can be removed and they can learn to visualise the numberline in order to answer the questions mentally.

Digit cards which include a decimal point can be used to further develop understanding and questions asked, such as 'Can you make me a decimal fraction that is greater than 3.4?' 'Make me a decimal fraction in between 3.1 and 6.7'. 'Make me the largest/smallest decimal fraction you can with just 4 cards' etc.

Children are encouraged to use the calculator to check or interpret answers. They need to be able to understand how to type amounts including decimals into a calculator and how to interpret decimal fractions on a calculator screen. For example, 'What could 5.2 on a calculator screen mean in terms of money, or length?' They should be able to add and subtract mentally to convert one decimal fraction to another, for example, know what to type into a calculator to change 4.7 to 4.9. Again, plenty of questions linking to the context of money or measurements need to be posed before more abstract questions are attempted.

It is important to ensure that your child understands the relationship between decimals and fractions (i.e. realises that decimals are a way of recording tenths and hundredths so, for example, 0.9=9/10 or 7.41= 7 ones, 4 tenths and one hundredth). They need to be able to both recognise and use this notation, initially in the context of money and measurements then, as their understanding grows, in more abstract contexts. You can help them to gain confidence with this by practising converting fractions to decimal fractions (e.g. 1/10 of a pound is 0.10 so 1/10=0.1 or 78 and 7/10 is the same as 78.7) and decimal fractions to fractions. Through the course of their work they should become aware of the equivalence between decimal and fraction forms of one half, one quarter, three quarters, one tenth and one hundredth (i.e. ½=0.5, ¼=0.25, 3/4=0.75, 1/10=0.1 and 1/100=0.01). By the end of the year they should be able to order sets of numbers (such as money or measurements) with one or two decimal places and position them on numberlines.

Fractions

As with decimals, it is important that children gain an understanding of fractions in contexts that they understand so work should initially be practical, involving the use of diagrams, shapes and objects. Children need to begin to appreciate the size of fractions in relation to each other. Some children find fractions difficult because the higher the denominator (bottom number) becomes, the smaller the fraction actually is. For example, one tenth is actually smaller than one fifth, even though ten is larger than five, so plenty of practical work is essential to secure understanding.

Help your child to identify fractions using diagrams and shapes drawn on squared paper with a certain fraction shaded. Initially unit fractions (such as ½, 1/3, 1/5) should be found then other simple fractions (such as 2/3, 5/6), making sure that they appreciate that a fraction always has to be an *equal* part. They obviously need to understand that the bottom number indicates the total number of parts that the whole is divided into, whilst the top number indicates a certain number out of those parts. As well as recognising fraction notation, also encourage your child to use it (e.g. 1/10 instead of one tenth) in order to build up their confidence. Shapes are useful for illustrating mixed fractions (such as 3½, 4 ¼) and giving an idea of their relative size.

Make sure that your child can find fractions of numbers as well as fractions of shapes. Work practically with small numbers to begin with and encourage them to appreciate the relationship between fractions and division, i.e. they should know that in order to find one tenth you must divide by 10. Your child should then use their knowledge of division to find fractions of numbers. For example, find 1/10 of 30; find 1/5 of 25. They should also be able to work out the answers to questions such as 'What fraction of £1 is 25p?' Again, use practical objects or draw diagrams or pictures to help make the link to division explicit. As with shapes, once they can find unit fractions (e.g. 1/5, 1/3, 1/8 etc) they should learn how to find other fractions such as 3/5, 2/3, 4/6. By working practically with shapes and smaller numbers, they can begin to realise that they can find larger fractions of quantities by finding how much one part is then multiplying by the amount needed. For example, to find 3/5, know that you find 1/5 then multiply by 3. Encourage your child to illustrate and explain this understanding through the use of diagrams and equipment and then to present their work pictorially in order to consolidate their understanding.

As with any numbers, counting forwards and backwards can help to develop understanding so encourage your child to count on and back in multiples of a half, a quarter, a fifth or a tenth. Help them to mark these multiples on numberlines which go beyond one so that they begin to understand the relationship between improper fractions and mixed fractions. For example, realise that 7/4 is the same as 1 and ¾. In Year Four, children need to be able to order a range of mixed fractions (e.g. 4½, 3¼ etc) and place them on numberlines so it can be useful to write a range on cards and ask your child to place them in order, explaining how they know which are bigger. Simple fractions could also be written onto cards and sorted according to whether they are larger or smaller than a half. The accompanying discussion in these types of activities can really help children to clarify their understanding. Always refer back to pictorial representations or diagrams until understanding is secure. Gradually your child should begin to gain an understanding of equivalent fractions involving halves, quarters, eighths, fifths, tenths, thirds and sixths, using diagrams to help them. It is a good idea to ask your child to investigate and find fractions that are equivalent to others. For example, 'Find fractions that are equivalent to one half or a quarter. What do you notice about them all? Can you predict any others?' 'Can you find any fractions that are equivalent to ¾? How do you know?'

There are a number of commercially available sets that allow children to put together fractions puzzles and see the link between, for example, fifths and tenths, quarters and eighths or thirds and sixths (i.e. that one is half the size of the other). I would highly recommend these if your child does find fractions difficult as they allow your child to use practical equipment to answer questions such as: 'How many quarters are equal to a half?', 'Which is bigger: a third or a sixth?' 'Find two fractions that are equal to one whole.' You could ask them to order unit fractions then check them using these practical sets. Shading, cutting and folding shapes on squared paper or numberlines can also help to establish the equivalence between halves and quarters, thirds and sixths or fifths and tenths.

Your child will also need to begin to investigate the equivalence between tenths and hundredths and link this to their understanding of decimals. For example, they may cut or shade a 10x10 square to show tenths and find that one tenth is equal to ten hundredths or find that 4 tenths and 3 hundredths is the same as 43 hundredths. They can begin to link this to their work on decimals and realise that one tenth can be expressed as 0.1 and 43 hundredths can be expressed as 0.43. By shading other amounts of a 10x10 shape to represent one half, one quarter and three quarters they can develop their understanding that ½ =0.5, ¼ =0.25 and ¾ =0.75. Numberlines can also be used and tenths can be marked on in both their fraction and decimal notation. The relationship between tenths and hundredths can then be illustrated again. For example, 3 tenths can be seen as 30 hundredths.

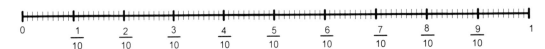

Figure 6.7

By the end of Year Four, children are expected to:

- Understand and use fraction notation.

92

- Have an understanding of simple fractions such as 1/3 or 4/5.
- Have an understanding of mixed fractions such as 4½, 6¾.
- Order mixed fractions and position them on numberlines.
- Through practical work, appreciate the relationship between
    - fifths and tenths, realising that 2/5 is equal to 4/10 etc.
    - thirds and sixths, realising that 2/3 is equal to 4/6 etc.
    - halves, quarters and eighths, realising that ½ is equal to 2/4 or 4/8 and ¾ is equal to 6/8.
- Identify fractions of shapes, initially unit fractions (1/4, 1/6) then other simple fractions (3/4, 7/8).
- Link fractions to division, i.e. know that to find 1/5 you must divide by 5.
- Find fractions of numbers (e.g. find 1/5 of 35, find 4/5 of 35), using practical equipment or diagrams to illustrate and explain the link to division and multiplication, until understanding is secure.
- Recognise the equivalence between fractions and decimals, for example, know that ½=0.5, ¼=0.25, 3/4=0.75, 1/10=0.1 and 1/100=0.01.
- Give pairs of fractions that total one, such as 4/5+1/5 etc.

Ratio and Proportion

Children need to become familiar with the terms 'in every' (which compare part to part) and 'for every' (which compares part to the whole) and discuss problems which describe the relationship between two quantities. They will begin to relate fractions to finding a proportion and estimate simple proportions. Through discussion, they will begin to understand the difference between these terms.

For example,
- 'On a fence, one post in every three is black. If there are 12 posts, how many will be black?' This means that 1/3 of the posts are black so 2/3 aren't black. It also means that there are twice as many posts that are not black. This can be worked out by drawing a quick diagram and marking on which posts are black.
- 'Kate has 12 sweets. She gives her friend one sweet for every three she takes. How many will her friend get?' This can be worked out practically or using pictures to begin with so that your child will come to realise that Kate will have ¾ of the total and her friend will have ¼. It also means that Kate will have 3 times as many as her friend.
- Children need to be familiar with the language 'in every', 'for every', 'as many as' …etc and be able to solve and talk about problems involving this vocabulary.
- They also need to be familiar with the language of 'scaling', for example: 'My brother is 3 years old and my sister is three times as old as him. How old is my sister?'

Rapid Recall of Addition and Subtraction Facts

Children cannot have rapid recall of every number fact. However, they should be able to work out a huge number of facts quickly in their heads by using those they already know to work out others quickly. In Year Four they should:

- Consolidate addition and subtraction bonds of all numbers from 1-20 (e.g. know all the number pairs that total 17 and their corresponding subtraction facts, such as 12+5=17 so 17-5=12).
- Consolidate understanding of using known number facts to work out new facts (e.g. 7+9=16; 70+90=160; 700+900=1600 etc). It is expected that children will be able to quickly work out sums and differences of multiples of 10, 100 or 1000 by using known facts as a starting point.
- Add three 2 digit multiples of ten (e.g. 40+50+20).
- Consolidate recall of pairs that total 100 (e.g. 56+44).
- Multiples of 50 that total 1000 (e.g. 150+850; 250+750 etc).

- Addition doubles from 1+1 to 50+50 (e.g. 48+48).
- Doubles of multiples of 10 from 10+10 to 500+500 (e.g. 390+390).
- Doubles of multiples of 100 from 100+100 to 5000+5000 (e.g.1900+1900=3800).
- Be able to double or halve any 2 digit number mentally through partitioning. (For example, to find double 48, double 40 then double 8 and mentally add 80+16). The doubles and halves of multiples of 10 and 100 should be found by applying this knowledge. (Because double 48 is 96, then double 480 is 960 and double 4800 is 9600).
- The corresponding halves for all doubles stated should be derived quickly.

It is expected that those facts learnt in previous years will still be recalled as they are still needed to help solve mental calculations quickly and easily.[13]

Mental Calculation Strategies (Addition and Subtraction)

The following strategies are expected to help children to use mental methods confidently and successfully. By now, children should be able to confidently choose appropriate strategies, depending upon the calculation.

- When adding by counting on, start with the larger number, unless another strategy is more appropriate.

- Add several small numbers, looking for pairs that make 10 or for doubles/near doubles and add these first. Also look for pairs that total 9 or 11 then adjust. With addition of multiples of 10, look for pairs that total 100 (e.g. 40+20+60. Add 40+60 first as it is a number bond to 100 and then add on the 20; 30+50+40. Add 30+40 first as it is a near double which can be worked out without counting. This will total 70. Then add the remaining 50 to 70 to reach 120). Looking for patterns can also help with addition of several numbers. For example, 6+5+7+6 may be spotted as equivalent to 6x4.

- Use awareness of near doubles (e.g. 36+35 is double 35+1; realise that 43+45 is the same as double 44).

- Partition (split) numbers, if appropriate, to make them easier to work with mentally.
  o Numbers may be partitioned according to place value, e.g. 324+57 can be solved by partitioning the second number and counting on, i.e. 324+50+7:

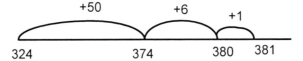

Figure 6.8

Or by partitioning both numbers: 300+20+50+4+7=300+70+11=381.
  o You will notice that in the example above single digits were also partitioned in order to use the bridging strategy (figure 6.8). Bridging is an important calculation strategy. It is used to speed up mental calculations as it allows children to make use of number bonds instead of counting in ones. Any multiple of 10, 100 or 1000 can be used as a 'bridge' (or stopping place) in order to calculate more efficiently.
  For example, in figure 6.8 to solve 324+57, 50 is added to reach 374 then 7 is added but, instead of counting on in ones from 374 to 381, the multiple of ten, in this case 380, is used as a 'bridge' or stopping place. The 7 is then

[13] See Rapid Recall section in previous chapters.

94

partitioned into 6+1 in order to make use of the multiple of ten and allow children to make use of their number bonds and work more efficiently. Obviously, it is extremely important that your child can mentally partition small numbers so that they can make use of 'bridges' in this way. This strategy also requires sound recall of number bonds to ten. (There are further examples and explanations of bridging in the Mental Calculation Strategies section of each chapter and also in the examples for mental calculations later in this chapter).

- Find the difference between numbers either side of a multiple of 10, 100 or 1000 by counting up (uses bridging strategy). For example, 92-89; 403-386; 4003-3994; children should add to reach the multiple of 10,100 or 1000 and use this as a bridge.

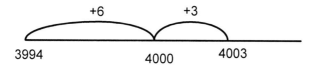

Figure 6.9

4003-3994: 6 more is needed to reach 4000 from 3994 then another 3 is needed to reach 4003 so 4003-3994=9.[14]

- Count on or back in repeated steps of 1, 10, 100 or 1000. This strategy is useful to help with addition and subtraction of larger numbers when children are still gaining confidence with the order of larger numbers and the relationships between them. However, once these are established, bridging should be encouraged where possible as it tends to make mental calculations quicker.

- Use compensation. This basically involves rounding a number up or down to make it easier to work with then adjusting it at the end. Children should be able to add a near multiple of ten (up to 100) to any 2 or 3 digit number quite confidently. For example, 567+38: add 40 instead of 38 then subtract 2. 4536+95: add 100 instead of 95 then subtract 5= 4631.

- Use understanding of the inverse relationship between addition and subtraction to help with calculations. So, if they know that, for example, 36+19=55 then they also know that 19+36=55, 55-19=36 and 55-36=19. Use this to check their own work.

- Approximate the answer first (e.g. 456+6733 is approximately 500+7000 so the answer will be a little smaller than 7500, because both numbers were rounded up). This helps children to gain an idea of the size of the actual answer and therefore realise when answers that have been calculated are incorrect.

- Check work by performing an equivalent calculation, the inverse operation, adding in a different order or using knowledge of odd and even numbers. For example, know that the difference between two even numbers is always even.

Addition

As in previous years, children need to:
- Understand that addition of positive whole numbers will make them larger (apart from when zero is added).

---

[14] See page 50 for a more detailed explanation of subtraction as counting on (complementary addition).

- Realise that addition can be done in any order e.g. 30+80=80+30, (30+80)+20= 30+(80+20).
- Understand that addition is the inverse of subtraction (i.e. it reverses it) and use it to check results.
- Respond rapidly to addition problems, explaining methods both orally and in writing.
- Record stages in calculations involving addition.
- Use mental methods to add 2 digit numbers; always check that a calculation cannot be done mentally before embarking upon a written method.

Subtraction

Children need to understand that subtraction:

- Makes numbers smaller (except for subtraction of zero which leaves a number unchanged).
- Cannot be done in any order, unlike addition (i.e. that 5-3 is not equal to 3-5).
- Is the inverse of addition (i.e. it reverses it).

Children need to consolidate their understanding of subtraction as:
- Taking away /counting back.
- Finding the difference.
- Complementary addition (counting up from the smaller to the larger number).

Children should use mental methods to subtract two 2 digit numbers, always checking that a calculation cannot be done mentally before embarking upon a written method.

Consolidating Addition and Subtraction

The revised framework states that, in Year Four, children should mentally add and subtract pairs of two digit numbers. Examples of appropriate calculations are given below. All of the following calculations build on the strategies taught in previous years and involve consolidation of previous strategies rather than learning new concepts. Consequently, children are expected to solve these types of calculations mentally and explain the method used both orally and in writing. If your child is unable to tackle any of the following mentally then examples of appropriate strategies which will help to develop mental methods are suggested. Make sure your child can confidently tackle the examples using pencil and paper to record or make 'jottings' if they lack confidence, and discuss their methods with them, encouraging them to explain their thinking. If you find that your child is not managing with the Year Four examples, it would be a good idea to refer back to the previous chapter and ensure that your child can solve the examples given there.

1.Continue to add and subtract 2 digit multiples of ten

E.g. 50+70=   60+__=90

Using known facts such as 5+7=12 so 50+70=120 is probably the most efficient strategy here. Partitioning numbers to make use of number bonds then using the 100 as a bridge may also be helpful. E.g.50+70; child may know that 70 is 50 and 20 and so choose to solve the calculation by saying 50+50=100. 100+20=120.

2.Add and subtract a pair of multiples of 100, crossing the thousands boundary

E.g. 800+400=   1400-600=

Again, children should be able to choose an appropriate strategy to use. It may be using known facts (8+4=12 so 800+400=1200) or bridging through 1000 and partitioning the 400: 800+200=1000. 1000+200=1200. Less confident children may solve this by counting on or back in hundreds. However, it is important that you also demonstrate the link to known facts to increase their awareness of this strategy.

3.Revise the addition and subtraction of multiples of ten to and from a 2 or 3 digit number, without crossing the hundred boundary

E.g. 456+ 40=   678- 60=   753+__=783

Knowledge of place value and patterns in numbers can be used here. 50+40=90 so 456+40=496. Place value arrows can help to reinforce this understanding.

4.Revise the addition of 2 and 3 digit numbers to multiples of 10, 100 or 1000

E.g. 40+25=   600+56=   500+156=   3000+458=

Some of these are effectively numbers that have been partitioned so children should be able to simply recombine them here. If your child finds examples such as 500+156 difficult then place value arrows should be used to demonstrate the value of the digits. Once the numbers are partitioned they can then be added and recombined. So 500+156 would be seen as 500+100+50+6. The hundreds can then be added and then the numbers recombined into 656.

5.Find what must be added to 2 or 3 digit numbers to make 100 (or next higher multiple of 100)

E.g. 72+__=100   567+__=600

In previous years this may have been calculated using a blank numberline to count on.

Figure 6.10

Children should now begin to visualise these stages on the numberline, using multiples of 10 or 100 as bridges, in order to solve these calculations mentally.

6.Find what to add to 4 digit multiples of 100 to make the next multiple of 1000

E.g. 4500+__=5000   7800+__=8000

Knowledge of number bonds and using known facts should help with this example.

7.Add a single digit to any 3 or 4 digit number, crossing the tens boundary

E.g. 458+9=   2376+7=   675+__=681

Children may again use known facts (8+9=17 so 458+9=467) but partitioning the single digit numbers then bridging is probably a more efficient strategy here. For example, 458+9; you know that 2 more is needed to reach 460 so split the 9 into 2 and 7 then say 458+2+7=467.

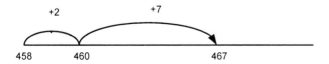

Figure 6.11

8.Subtract a single digit from a multiple of 100 or 1000

E.g.300-5=   4000-6=

Use knowledge of number bonds to work this out. A secure grasp of number is required; your child must realise that 300 can be split mentally into 200+100 then the 5 subtracted or the 4000 partitioned into 3000+900+100 then the 6 subtracted. These types of calculations often pose a difficulty for children if their knowledge of number order and place value is insecure. Plenty of work involving subtracting from one hundred or one thousand can help with this. Once children can confidently subtract a single digit from one hundred or one thousand, they can begin to subtract from two hundred or two thousand and start to see the pattern that emerges.

9.Subtract a single digit from a 3 or 4 digit number, crossing the tens boundary

E.g. 664-8=   7842-4=

Again, partitioning the single digit number and bridging through the last multiple of ten would be an efficient mental strategy. So 664-8; split the 8 into 4 and 4 then say 664-4-4=656.

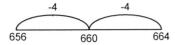

Figure 6.12

10.Find a small difference between numbers lying either side of a multiple of 1000, by counting up

E.g. 5004-4992=

Use the multiple of 1000 as a bridge, then use number bonds to reach 5000 from 4992(+8) then add 4 more to reach 5004. Total added 12.

11.Add or subtract any pair of 2 digit numbers, including crossing the tens boundary

E.g. 56+47=   98-45=   45+__=89

Children should choose the most efficient strategy to use mentally. The process of partitioning the numbers is best followed for some calculations. (For example, 56+47= 50+40+6+7= 90+13=103). In other cases, it is best to round the number to the nearest multiple of ten then adjust (compensation). For some subtractions the tens may be subtracted then the ones (e.g. 53-12=53-10-2=41); in other cases, it is quicker to count up from the lower to the higher number (e.g. 53-48; 2 more from 48 will reach 50 then 3 more will reach 53 so the total difference is 5). It is important to help your child to make decisions about which strategy is the most appropriate through discussion as they are working.

12.Mentally add and subtract 3 digit multiples of 10

E.g. 570+250   620-380   610-__=240   ___-370=240

This was in the original Numeracy Framework for Year Five. However, if children are able to add and subtract 2 digit numbers and have a secure understanding of place value, they should be able to use this knowledge to solve calculations such as these. Because 57+25=82 we know that 570+250=820. A range of other strategies could also be used depending upon the numbers involved. For example, to solve 570+250, some children may partition the numbers, saying 500+250=750 then partition 70 into 50 and 20 and add to 750 to reach 820.

Compensation could be used, 600+250=850 then subtract the 30 to reach 820. Any of these mental methods would be valid and many others too.

Developing Written Methods (Addition)

A general introduction to written methods is given in the previous chapter[15] which explains the reasons for the changes to the teaching of written methods and the significance of the expanded methods now taught.

By the end of Year Four it is expected that the majority of children will be using efficient written methods for both addition and subtraction when working with numbers which are not easy to work with mentally. Vertical written methods for working with 2 and 3 digit numbers may be introduced in Year Three. In Year Four, children are also expected to use written methods for working with decimals in the context of money. For some children, these methods may be the standard compact methods that many parents are familiar with from their own schooldays. Certainly, if children have a secure understanding of expanded methods, they should be introduced to the standard method which is more efficient. However, this must only be introduced if children are ready and fully understand it. There is no actual requirement for children to use the standard method until much later. They are simply expected to use an efficient written method.

Throughout the year, the emphasis will be on simplifying and reducing calculations so that only the stages in calculations that are necessary for the children are recorded. This will allow children to use the expanded methods to record written calculations in the most efficient way possible.

Expanded Methods

A range of expanded methods have been used in primary schools since the introduction of the Numeracy Strategy and I have included all those that were in the original document, as well as those added by the revised framework. However, to try to use them all would be confusing for children; schools tend to teach certain methods and use others merely to clarify understanding or to offer an alternative to children who find the main method difficult.

Partitioning

The methods for addition which involve partitioning were introduced and discussed in Year Three. The method below (figure 6.13) was introduced in the revised framework and is useful for children just gaining confidence with vertical methods as it is very similar to the jottings that accompanied earlier mental methods where both numbers were partitioned.

$$367 + 285 = \qquad \begin{array}{rrr} 300 + & 60 + & 7 \\ 200 + & 80 + & 5 + \\ \hline 500 + & 140 + & 12 \quad = 652 \end{array}$$

Figure 6.13

The method shown in figure 6.14 is a more compact method when using large numbers, although it is not necessarily as easy to understand as the method above. Both methods link to mental methods of addition and ensure an understanding of the size of the numbers that are involved.

---

[15] See page 73.

```
3   6   7
2   8   5   +
_____
5   0   0  (add hundreds)
1   4   0  (add tens)
    1   2  (add ones)
_____
6   5   2  (TOTAL)
```

Figure 6.14

It is important that children are involved in discussion about the order in which the digits are added. In the examples above, the most significant (largest) digits are added first, in this case the hundreds. However, there needs to be some discussion of whether hundreds or units should be added first and if it makes any difference to the final total. At some point they need to realise that although the hundreds are added first because this is the way they will have been used to adding mentally, adding the units will give exactly the same answer. Children need to experience and feel confident with both ways of working and eventually be encouraged to always add the units first, in preparation for the standard written method which always starts with the units.

```
3   5   8
    7   3   +
_____
    1   1  (add ones)
1   2   0  (add tens)
3   0   0  (add hundreds)
_____
4   3   1  (TOTAL)
```

Figure 6.15

Expanded written methods may also be required when adding decimals in the context of money. It is important to know that the decimal points must line up beneath each other, especially if mixed amounts (e.g. £1.67 + 94p) are to be added.

```
£ 1 . 6   7
  0 . 9   4   +
_____
  1 . 0   0  (add pounds/units)
  1 . 5   0  (add ten pences/tenths)
  0 . 1   1  (add one pences/hundredths)
_____
  2 . 6   1  (TOTAL)
```

Figure 6.16

Compensation.

The following is an extension of the mental calculation strategy of compensation. This builds upon the earlier mental strategies employed in previous years but records them in a vertical format. This vertical recording could be recorded alongside numberline work initially to aid understanding of the vertical method.

```
    7   5   4
        8   6   -
    ─────────
    8   5   4   (100 added instead of 86)
-       1   4   (subtract 14 as 86 is14 less than 100)
    ─────────
    8   4   0   (TOTAL)
    ─────────
```

Figure 6.17

```
£   1 . 6   7
    0 · 9   4   +
    ─────────
    2 · 6   7   (add £1.00)
    0 · 0   6   (subtract £0.06 as £1.00 is £0.06 more than £0.94)
    ─────────
    2 · 6   1   (TOTAL)
    ─────────
```

Figure 6.18

Compensation in its vertical layout is included in the original Numeracy Framework. However, there appears to be no mention of it in the *vertical* format in the revised framework. I would imagine that this may be because, although it is appropriate in certain cases, it is not the most appropriate method for the majority of calculations and therefore is not used often as an expanded method. It tends to be used more often as a mental strategy.

Standard Compact Method

If your child is very confident with the expanded method they may be introduced to the compact written method, which is the one most of us will be familiar with from our school days and will probably employ when we carry out calculations.

```
    3   5   8
        7   3   +
    ─────────
    4   3   1
    ─────────
    1   1
```

Figure 6.19

```
£   1 . 6   7
    0 · 9   4   +
    ─────────
    2 · 6   1
    ─────────
    1   1
```

Figure 6.20

101

Bear in mind that this is only to be used if your child has a full understanding of expanded methods and can use the compact method with consistent accuracy. If errors begin to occur when using compact methods, it indicates lack of readiness and expanded methods should be returned to with a view to making them as efficient as possible. However, if your child is very confident with the expanded methods and has a good grasp of the place value of the numbers then it is a good idea to encourage the use of the standard method as it is quicker and more efficient.

Developing Written Methods (Subtraction)

In Year Three children may have been introduced to a vertical method which corresponds to counting back on a numberline. The following method corresponds to counting up on a numberline.

Counting Up (Complementary Addition)

For children who struggle with subtraction, this is sometimes an easier method to use, as it involves counting up rather than counting back. It would be demonstrated alongside a numberline to begin with (see Figure 6.21).

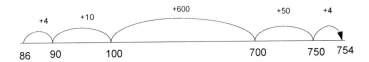

Figure 6.21

```
7   5   4
    8   6   -
        4  (to reach 90)
    1   0  (to reach 100)
6   0   0  (to reach 700)
    5   0  (to reach 750)
        4  (to reach 754)
6   6   8  (Total counted on from 86 to 754)
```

Figure 6.22

As children become more confident, they are encouraged to make the calculations shorter and more efficient by reducing the steps if possible, and solving more of the calculation mentally as shown:

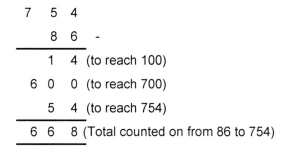

```
7   5   4
    8   6   -
    1   4   (to reach 100)
6   0   0   (to reach 700)
    5   4   (to reach 754)
6   6   8   (Total counted on from 86 to 754)
```

Figure 6.23

It is important that children record the steps that they need to solve the calculation so do not push your child to reduce the calculation unless they can comfortably do so. However, calculations need to be as efficient as possible so discourage them from spending unnecessary time recording stages in calculations if they are able to solve them mentally.

Compensation

The original Numeracy Strategy also showed the vertical recording for compensation as follows:

```
7   5   4
    8   6   -
6   5   4   (take off 100 instead of 86)
    1   4   (because 100 is 14 more than 86)
6   6   8
```

Figure 6.24

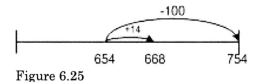

Figure 6.25

As mentioned previously, this vertical recording of compensation is not mentioned in the revised framework and does now appear to be a lower profile written strategy.

Partitioning (Decomposition)

This method can be quite difficult for children to grasp. The numbers are partitioned (as in the expanded method for addition) and simply subtracted. However, if there are not enough in the tens column to subtract what is required, the number is partitioned again, this time moving a hundred into the tens. (If there are not enough in the units to complete the subtraction then the number may be partitioned again to move a ten into the units). A good understanding of larger numbers and the ways in which they can be partitioned is required to use this method confidently. Exploration of ways to partition 3 digit numbers by moving hundreds into tens and tens into ones should be carried out alongside this type of calculation if your child finds it difficult. (See Year Three).

103

For example,

```
 7 5 4              7 0 0 +   5 0 +   4
   8 6  -                     80 +   6  -
 _____           _____
```

Figure 6.26

This subtraction is not easy to solve so partition 754 as 700+40+14 instead.

```
                40        14
 7 0 0 +      5̶0̶ +      4
             80 +      6  -
 _____
```

Figure 6.27

This still cannot be solved so partition it again into 600+140+14. Now the subtraction can be solved.

```
   600        140        14
                40
 7̶0̶0̶ +      5̶0̶ +      4
             80 +      6  -
 _____
 6 0 0 +     60 +      8   =  6 6 8
```

Figure 6.28

In the case of a zero, it can be rather more complicated as the number has to be partitioned in two stages.

E.g. 602-365

```
   500       90          12
             1̶0̶0̶
 6̶0̶0̶ +     0̶ +        2
 3 0 0 +     60 +        5  -
```

Figure 6.29

This method involves partitioning the numbers in such a way as to ensure that there are enough in the tens and hundreds to solve the calculation. It allows children to see what is actually happening in the standard compact method when numbers are 'borrowed' from the tens or hundreds, thus eventually leading to a full understanding of the standard method. [16]

```
       ¹4
   7̶   5̶ ¹4
         8  6   -
 _____
   6   6  8
```

Figure 6.30

---

[16] The order of difficulty for calculations involving decomposition is shown in the relevant section in Year Three.

Multiplication

It is important to understand:

- Multiplying by one leaves a number unchanged.
- Multiplying by zero=0.
- Multiplication as repeated addition e.g. 72x3=72+72+72 or 72 sets of 3.
- Multiplication can be done in any order, so 9x2=2x9; (4x5) x3=4x(5x3).
- Partitioning can be useful to help solve multiplication problems as well as addition and subtraction.
- Multiplication is the inverse of division (it reverses division) and use this to check work.

As in previous years, children will be expected to count on and back in steps of constant size from different starting points and to continue or fill in the missing numbers in number sequences, looking for and explaining patterns in them. Number sequences may now include decimals and negative numbers.

Rapid Recall (Multiplication and Division)

- Know by heart the times tables facts up to 10x10; quickly work out the corresponding division facts (so if 6x7=42 then 42 divided by 7 =6).
- Derive quickly related facts using knowledge of place value (4x5=20 so 40x5=200).
- Recognise multiples of all numbers to 10 up to the tenth multiple.
- Know doubles of all numbers 1 to 50 (i.e. 12+12 or 49+49).
- Know doubles of all multiples of ten to 500 (i.e. any number you would say when counting in tens, such as 120+120 or 490+490).
- Know doubles of all multiples of 100 up to 500 (i.e.any number you would say when counting in hundreds, such as 200+200,400+400 or 500+500).
- Know all the corresponding halves for the doubles they know.
- Be aware of the effect of multiplying or dividing by 10 or 100.

Mental Calculation Strategies (Multiplication)

Although there appears to be a huge range of strategies to learn here, most are actually based on the same concept and simply require an awareness of the relationship between doubles/halves and how they can be used to allow us to work flexibly during multiplication.
- Know that to multiply by 4 you can double it then double it again.
- Know that to multiply by 5 you can multiply by ten then halve it.
- Know that to multiply by 20 you can multiply by 10 then double it.
- Know that to multiply by 8, you can multiply by 4 then double it.
- Work out some multiples of 15 by doubling. Combine some of the facts gained to find new facts. E.g. add 2x15 to 4x15 to find 6x15.
- Find ¼ by halving then halving again or 1/8 by halving a quarter.
- Use combination of multiplication facts:
  - Work out x6 by x4 then x2 added together.
  - Work out x9 by x10 then subtracting the number.
  - Work out x11 by x10 then adding the number.
- Partition numbers in order to multiply a 2 digit number by a single digit number. E.g. 24x4= (20x4)+(4x4). Also partition when doubling and halving: double 45= double 40 and double 5 etc.
- Use compensation e.g. 49x5= (50x5)-5.
- Continue to use and explore the inverse relationship between multiplication and division, realising that knowing one fact means that another three facts can be gained. For example, know that because 12x9=108, then 9x12=108, 108 divided by 12=9 and 108 divided by 9=12.

Using rhyming words to help your child to learn a tricky fact can be useful, particularly for children who dislike rote learning. For example, to learn 6x8, think of something to rhyme with 6 (kicks) and something to rhyme with 8 (gate), then something to rhyme with 48 (naughty gate). Then make up a silly story with your child about a character who kicks a gate and says 'naughty gate' – six (times) eight, forty eight: kicks gate, naughty gate! The same rhyming word can be used for a certain number so every fact involving the eight times table could include a gate. For some children this really does help.

Division

Children need to:
- Understand division as both sharing and as grouping. (This is extremely important as mental methods, expanded and compact written methods involve grouping rather than sharing).
- Know that dividing by one leaves a number unchanged.
- Know that division is the inverse of multiplication (and should use this knowledge to check calculations). Demonstrating division on a numberline can help to make the link between multiplication and division more explicit.
- Begin to relate division to fractions (i.e. know that to find ¼ you need to divide by 4).
- Be able to give whole number remainders and decide whether to round up or down when a remainder occurs in a problem.

Multiplying and Dividing by Ten and One Hundred

In Year Four children consolidate their understanding of multiplying and dividing numbers by 10. Initially through practical work, they will have gained an understanding that multiplying a number by ten will cause the digits in that number to move one place to the left and that dividing by ten will cause the digits in a number to move one place to the right. (If your child does not have a secure grasp of this concept, more details are given about this in Year Three). Moving on from this, children will also learn how to multiply and divide whole numbers up to 1000 by 100. (Bear in mind, numbers to be multiplied or divided by 10 or 100 should result in whole number answers). Problems involving multiplication/division by 10 or 100 often involve 'scaling'. More details about this concept can be found in the section on 'Ratio and Proportion'.

Children should be able to use known number facts and their understanding of place value to multiply and divide mentally. This does not mean that they must have instant recall of the facts below but they must be able to work them out quickly in their heads. If they are not quite at this stage, then encourage 'jottings' and informal recordings which support mental methods, rather than acting as a substitute for them. They should also be able to explain their methods in writing.

The Numeracy Frameworks for Year Four (original and revised) include the following calculations for multiplication and division.

1. Multiply a 2 or 3 digit number by 10 or 100

E.g. 327x10   54 x100   82x___=8200

2. Divide 4 digit multiples of 1000 by 10 or 100

E.g. 8000 divided by 10   5000 divided by 100   6000 divided by ___=6
Find 1/10 or 1/100 of 4000.

3.Double any multiple of 5 up to 100

E.g. double 75: double 70=140, double 5=10 so double 75=150.

4.Halve any multiple of 10 to 200

E.g. halve 130: halve 100=50, halve 30=15 so halve 130=65.

5.Consolidate multiplication of a 2 digit multiple of ten by 2, 3, 4, 5 or 10. Begin to multiply 6,7,8 and 9

E.g. 20x3=__   70x6=__   500=10x__   __= (2x60) +4

Using known facts and awareness of pattern in number is required here. For example, 2x3=6 so 20x3=60. The revised framework has emphasised the need to work with the 6, 7, 8 and 9 times tables in Year Four and expects rapid recall of these tables. Once these are known there is no reason not to carry out these calculations confidently when multiplying by any single digit.

6.Multiply a 2 digit number by 2, 3, 4 or 5 crossing tens boundary

E.g. 13x5 =   18x__=54

Using flexible strategies such as 13x10=130 so 13x5= halve 130 =65 or approximating first to gain an idea of the answer should be used (e.g. 20x3=60 so 18x3=54). Now that the revised framework has stated that times tables should be memorised earlier, these types of calculations could also include multiplication by 6, 7, 8 or 9.

For most of the above examples, children should be able to approximate their answer first using what they know about numbers, before using known number facts, awareness of patterns and partitioning etc to work out the final answer.

Developing Written Methods (Multiplication)

In Year Four, children will consolidate their understanding of multiplying 2 digit numbers by single digit numbers (by partitioning and multiplying the tens first). They may begin by recording their working out in an informal way such as

35x5

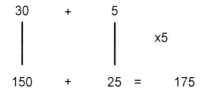

Figure 6.31

They may simply record the stages: 35x5=(30+5) x5
                                    (30x5)+(5x5)
                                    150+25=175.

However, the grid method is probably the most commonly taught expanded method for short multiplication in schools. It links directly to mental methods, allowing children to partition the numbers then use their understanding of pattern in number and tables facts to solve the calculation. It can be recorded in one of two ways. The following layout was suggested by the original Numeracy Framework and is useful as it does link with the layout for division demonstrated at a later stage.

107

E.g. 58x8

x	50	8
8	400	64

Figure 6.32

Total= 400+64=464

The revised framework suggested that it may be advantageous to place the larger number in the left hand column of the grid so that it is easier to add the partial products. This would be set out as follows:

x	8
5 0	4 0 0
8	6 4
(Total)	4 6 4

Figure 6.33

Both of these methods allow children to see what is actually involved when multiplying 2 digit numbers and lead to a greater understanding of what they are doing.

The revised framework then suggests an extra stage in the grid method which involves moving the number being multiplied to the top of the grid as shown in the example below.

	5 0 + 8
x	8
	4 0 0
	6 4
(Total)	4 6 4

Figure 6.34

This helps to prepare children for the expanded methods of short multiplication, which are recorded in columns.

```
  50 + 8                58
      8 x                8 x
  4 0 0  (50x8)      4 0 0  (50x8)
    6 4  (8x8)         6 4  (8x8)
  4 6 4  (Total)     4 6 4  (Total)
```

Figure 6.35

These methods encourage the children to gain an idea of the size of the numbers with which they are working so that they are more likely to recognise an answer that is far too high or too low.

Developing Written Methods (Division)

In Year Four, children are expected to use expanded methods to divide two digit numbers by single digit numbers. As with other calculations, children are encouraged to approximate first to gain an idea of the size of their answer. The following methods show written methods that can accompany the development of mental methods for division. They involve grouping rather than sharing, using tables facts to make larger groups.

Partitioning

Numbers can be partitioned and divided separately (then recombined) just as in multiplication calculations they can be partitioned and multiplied separately. Partitioning simply makes the number more manageable and allows children to make use of the facts that they already know. To find the best way to partition, it is usually a good idea to use multiples of ten as a starting point, and ask the question 'What is the highest multiple of ten that the divisor will divide into?'

For example, 72 divided by 5 =___. An approximate answer can be worked out as follows: We know that 50 divided by 5 = 10, therefore 100 divided by 5 = 20, so the answer should be between 10 and 20.

In order to divide, children are encouraged to use a multiple of the divisor (in the above example 5) to partition the dividend (number being divided). To work out how to partition the number, children are asked to use multiples of ten as a starting point.

In the above example, by using the fact that 10x5=50 (so therefore there are at least 10 fives in 72) a child can work out that it would be sensible to partition 72 into 50+22.
We know that there are 10 fives in 50. The next stage is to work out how many fives are in 22.

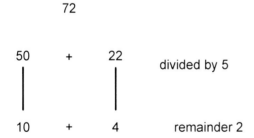

Figure 6.36

There are 10 fives in 50 and 4 fives in 22 with a remainder of 2 so (50+22) divided by 5=10+4 remainder 2.

72 divided by 5 = (50+22) divided by 5
                  (50 divided by 5) + (22 divided by 5)
                    10+4 R2= 14 R2
Obviously numbers that divide without remainders would be used initially until children gain confidence with the process.

Grid Method

Another method to record this process has been introduced by the revised framework and is shown below. This method links to the grid method of multiplication and shows the link between multiplication and division. As in other methods, the number should be partitioned, using the knowledge of multiples of ten as a starting point. In the example below, 96 divided by 6, the children should be asked,' How many sixes are in 60?' and then 'How many sixes in 36?' It would probably be wise, if introducing this method, to demonstrate it initially using a numberline as well.

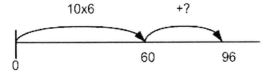

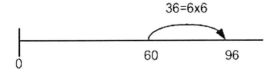

Figure 6.37

96 divided by 6:

X		
6	60	36

Figure 6.38

Children can use their multiplication facts to solve the division and fill in the grid as shown below.

X	10	6
6	60	36

Figure 6.39

110

<u>Using Multiples of the Divisor</u>

This method is sometimes called 'chunking' and it involves repeated subtraction. As children are developing their mental methods for division, they may choose to subtract repeatedly as an alternative method to partitioning. This method can be particularly useful for strengthening mental division strategies with larger numbers as it involves using known facts and place value to find the appropriate numbers to subtract. E.g. 96 divided by 4:

```
        9   6
  -     4   0    (10x4)
        ‾‾‾‾‾‾
        5   6
  -     4   0    (10x4)
        ‾‾‾‾‾‾
        1   6
  -     1   6    (4x4)
        ‾‾‾‾‾‾
            0
```

Figure 6.40

24 groups of 4 were subtracted from 96 so the answer is 24. This process should gradually be made more efficient by subtracting higher multiples of ten (so instead of 10x4, 20x4) as children become more confident.

<u>Short Division</u>

Once children confidently understand how to divide mentally using one of these methods as a starting point, they can move onto short division which is a more compact means of recording the mental method of partitioning. It can be introduced when children are confident with multiplication and division facts and can subtract multiples of ten confidently. They must also have a secure grasp of place value and partitioning. It is suggested for use in the revised strategy at the end of Year Four or the beginning of Year Five.

96 divided by 6 (approximately 100 divided by 5= 20)
This is recorded as follows:

$$6 \overline{\smash{\big)}\, 60+36} \quad \begin{array}{c} 10+\ 6 \end{array}$$

Figure 6.41

To find out how to partition the number 96, children have to ask themselves 'What is the highest multiple of 6 that is also a multiple of 10 (but less than 96)?' When children have a firm grasp of this method, it can eventually (in later years) be recorded in the more compact form:

$$6 \overline{\smash{\big)}\, 9\,^3 6} \quad \begin{array}{c} 1\ \ 6 \end{array}$$

Figure 6.42

Although this looks identical to short division as taught prior to the Numeracy Strategy, the discussion surrounding the calculation is very different as it still links closely to mental methods. The question to be asked is 'How many sixes divide into 90 so the answer is a multiple of 10?' This gives the answer 10 sixes or 60 with 30 remaining. This 30 is added to the 6 to make 36. Then the question 'What is 36 divided by 6?' is asked to complete the calculation. It is important to use this language so that the approach remains consistent with mental methods.

To summarise, in Year Four children continue to develop their mental methods by using the mental calculation strategies taught in previous years with larger positive numbers, negative numbers, decimals and fractions. Where the numbers are not easy to manage mentally, written methods are used. Making decisions is an important part of mathematics so children are encouraged not only to choose appropriate strategies and operations to solve problems, but also to decide whether to solve problems mentally or using pencil and paper and to make decisions about what exactly to record and how. This is one of the reasons that discussion and explanation of their work is so important. By talking their ideas and methods through, they can clarify their thoughts and gain feedback about ideas, thus helping to develop them further and deepen their understanding. Whilst in earlier years, discussion will have centred around explanation of methods, now the discussion will also start to move towards ways of applying what they know in various contexts.

More is now expected in terms of independence: children should now be making approximations of their answers routinely before they begin a calculation in order to feel confident that their solution is the correct one. They are also expected to check their work using appropriate strategies. They may, for example, choose to use the inverse operation (so to check 135 divided by 3=45, they may say 45x3=135). They may choose to do an equivalent calculation. For example, 76+39 may initially be solved using compensation: 76+ 40-1=115. It may then be checked by partitioning: 76+39= 70+30+6+9; 100+ 15=115.

Mental recall of facts is still required; indeed many mental calculation strategies rely upon this. This can be a difficult area for some children as it requires not only the initial learning of facts but also the ability to memorise them at a later date. To do this they have to be frequently revised, either at home or at school (preferably both!) and used in the context of their work. If you find that your child lacks confidence with the mental recall of facts for Year Four, make sure that they are secure with the facts needed for Years Two and Three. It is far better for them to learn these first, as once they have these, their confidence will increase and they will find it far easier to learn the new facts needed. If you can think up a silly rhyme or story to illustrate or explain a fact so much the better! Encourage them to use the facts they do know constantly during calculations. The more often they are used, the more embedded they will become. Finally, try to make learning fun. Short sessions are more constructive than longer ones and games which involve using cards or practical equipment are more likely to be enjoyed than pages of sums.

Chapter 7 :   Year 5

In Year Five, children consolidate and extend much of the work from Year Four. The same mental calculation strategies should be employed but in more difficult contexts and written methods should generally become more efficient and compact. Children will be expected to become more independent both in terms of making approximations before beginning written calculations and by checking their work after calculations. The written methods already introduced will be developed further and made as compact as possible. You will notice that some of the sections in this chapter are quite brief and consist mainly of lists detailing what children should know. This is purely to avoid needless repetition. In these sections refer back to the relevant parts in the previous chapter on Year Four for more detailed explanations and ideas for activities.

<u>Place Value</u>

Children in Year Five need to extend their confidence with place value (thousands, hundreds, tens and units) to the extent that they can work confidently with 4 and 5 digit numbers and can recognise 6 and 7 digit numbers. In order to do this, they must consolidate the place value work from Year Four and then extend their understanding by practising the same types of activities based around reading, writing, partitioning and ordering with 5 digit numbers. E.g. 'What number is halfway between 27,400 and 28,000?' Numberlines may be used to illustrate or consolidate this work. As in previous years, uncalibrated numberlines may also be used to estimate the position of numbers. E.g. 'What number is this? Explain how you know.'

Figure 7.1

It is worth reading the section on Place Value[17] in Year Four for ideas of how to help your child develop their understanding. Once they are confident with the reading, writing, partitioning [18] and ordering of 4 digit numbers, they can extend their understanding by carrying out exactly the same types of activities with larger numbers. Numberlines with divisions may be used to place unknown numbers but, as children gain confidence, empty lines such as the one above can also be used to help children further develop their ability to approximate.

By the end of the year children should be able to:

- Place 4 and 5 digit numbers onto a calculator display and then change one number to another simply by adding or subtracting. For example, they should be able to work out what must be added to change 75,567 to 95,567 (answer: 20,000).
- Identify which is the larger of two numbers (e.g. 23,647 and 23,467).
- State which number is half way between two 5 digit numbers.
- Round 2, 3 and 4 digit numbers to the nearest ten, hundred or thousand.

---

[17] See page 87.
[18] See explanations in 'Mental Calculation Strategies', page 118. Also explained in glossary.

<u>Decimals</u>

In Year Five children consolidate their understanding of decimals up to 2 decimal places. Although the work does gradually become more abstract, make sure that decimals are still linked back to the contexts of money and measurement as they were in Year Four and that practical equipment such as coins and numberlines are available for your child to use to demonstrate or explain their understanding.

It is important that your child has a secure grasp of what each digit of a decimal number (with up to 2 decimal places) represents. For example, they should know that the number 5.93 is made up of 5 ones, 9 tenths and 3 hundredths. They will be expected to order, partition and recombine decimals numbers, linking their work back to the context of money or measurement.[19]
They will also be expected to:

- Count on and back from any given number in decimal steps, including extending back into negative numbers e.g. count on/back in steps of 0.4 or 0.03.
- Begin to predict missing numbers in a decimal sequence and describe or explain the pattern.
- Order decimals from largest to smallest or vice versa and position them on numberlines. E.g. 'Where would you place 4.5?'

4.2   4.9

Figure 7.2

- State the decimal number before or after another (e.g. state the decimal number before 7.8) and state decimals between 2 others (e.g. give a decimal fraction between 5.6 and 5.8).
- Express fractions as decimal fractions and vice versa. So, for example, they should be able to express 45/100 as 0.45 or 65 2/100 as 65.02 etc.
- Use numberlines to count in tenths and hundredths from various starting points and understand the relationship between tenths and hundredths.
- Partition decimals using both fraction and decimal notation. For example, partition 6.38 as 6+0.3+0.08 and as 6+ 3/10+ 8/100.
- State what to add and subtract to change one decimal to another. For example, know that to change 46.82 to 46.52 then 0.3 must be subtracted.
- Round decimals with up to two places to the nearest whole number.

It is expected in Year Five that children will use efficient written methods to add or subtract a pair of decimals that are too difficult to work with mentally. However, they are still expected to use mental methods in order to approximate their answer and to make use of known facts to derive facts about decimal numbers (e.g. 56+45=101 so 5.6+4.5=10.1). They should also be aware of the effect of multiplying and dividing decimals by 10, 100 or 1000.

Decimals are often used in problems involving money or measures so calculations should be posed in the form of a problem whenever possible. Many problems require children to convert between units (e.g. pounds to pence or centimetres to metres or millimetres) so a firm grasp of the units involved is also required. Calculators are sometimes used to solve problems so it is essential that children are able to interpret decimal fractions (such as £3.50) appropriately

---

[19] The activities shown on page 89 will give ideas to help children practise this.

on a calculator display and can represent decimal amounts (when working with mixed units such as £2.29, £1.65 and 45p, for example) on a calculator.

Fractions

Children need to understand all that has been covered in Year Four.[20] It is important to continue to consolidate this understanding through the use of diagrams and practical equipment. They should continue to establish relationships between common fractions through their practical experience. So, for example, know that one seventh is less than one sixth or one sixth is half of one third. They should also be able to answer simple questions such as 'Which fractions are between three tenths and seven tenths?' 'Which fractions are less than ½?'

Children will also be expected to:
- Count on in steps of different sizes, e.g. count on in quarters extending past one whole, showing an understanding of mixed fractions.
- Order mixed fractions (such as 1 and ¾, 4 and 5/6) then improper fractions (such as 7/4, 9/6), according to their size etc.
- Compare mixed fractions and improper fractions, initially by placing on numberlines, to establish equivalences, e.g. 1 and ¾ =7/4.
- Explain how to turn a mixed fraction into an improper fraction, drawing diagrams to support work if necessary. Answer questions such as: 'How many quarters in 3 wholes? In 4 ½?' Discuss how to work this out without a numberline.
- Express a smaller number as a fraction of a larger group, e.g. 3 out of 4 is the same as ¾.
- Explore and explain equivalent fractions, so find out that 6/18=12/36 or 4 and 5/6= 29/6 etc. Initially work practically, comparing shapes on squared paper or folding strips. For example, fold a strip of 20 squares into quarters and colour three quarters to show that 15 is three quarters of 20. Find other equivalences in similar ways and record results, looking for and explaining patterns.
- Find and explain equivalent fractions using numberlines. (For example, on a numberline marked in twelfths, mark 1/3 and 5/6). Also, work out families of equivalent fractions by scaling up. (For example, ¾=6/8, 9/12 etc). Recognise that the numerator and the denominator are both multiplied (or divided by) by the same number to find an equivalent fraction.
- Through their work, come to recognise patterns in fractions, e.g. ½= 2/4= 3/6 etc (i.e. recognise that the numerator/top number is half of the denominator/bottom number), 1/3= 2/6= 3/9=5/15 (i.e. begin to recognise that the numerator is one third of the denominator).
- Understand the equivalence between fraction and decimal forms, expressing tenths and hundredths as fractions and decimals and explaining how they know. For example, 'What is the decimal equivalent of 3/10? Or 37/100?' Use numberlines to begin with. For example, mark tenths on in both fraction and decimal notation. This work can then lead your child to investigate other equivalences, such as a fifth is equal to 0.2.
- Gain familiarity with hundredths (through use of diagrams/ equipment) and recognise their equivalences e.g. know that 10/100=1/10, 25/100=1/4, 75/100=3/4.
- Understand the relationship between fractions and division i.e. know that 3 divided by 4 is actually ¾; 16/4 is the same as 16 divided by 4; use division and multiplication to find fractions of quantities.
- Begin to use their knowledge of the relationships between numbers when working with fractions, so find ¼ by halving a half or find 1/6 by halving a third.
- Solve problems involving fractions, deciding whether or not it would be appropriate to use a calculator. If using a calculator, children need to be confident about what to

---

[20] See pages 91 for examples of activities.

115

enter. For example, 'I have £194.50 and I want to spend 5/12 of it on a bike. How much will I have left?'

- Use calculators to express fractions as decimal amounts. For example, express ¾ or 3/20 as a decimal. Your child needs to realise that ¾ is equivalent to 3 divided by 4.
- Consolidate understanding of the equivalence between fractions, decimals and percentages.

Children are encouraged to explain their work and relate back to the use of diagrams and practical equipment in order to illustrate and explain their understanding. For example, they may fold or cut a shape made up of 100 squares to show how 50/100 is equivalent to ½.

Percentages

Percentages are introduced in Year Five. Practical resources such as money and 10x10 squared grids are used initially to develop understanding. By shading parts of a grid, for example, the relationship between fractions, decimals and percentages can be demonstrated. Children need to:

- Recognise the symbol %.
- Understand percentage as the number of parts in every hundred.
- State a fraction of 100 as a percentage e.g. 56/100=56%.
- Know one whole is 100%, ½ is 50%, ¼ is 25%, ¾ is 75%, 1/10 is 10%.
- Express tenths and hundredths as percentages. Know 10% =0.1, 20%=0.2 because they are tenths of the whole amount. Similarly, 1% is 1/100 of the whole amount so it is 0.01.
- Be able to say what percentage of a shape is coloured.
- Find percentages of different amounts. E.g. find 5%, 20% or 30% by finding 10% then halving, doubling etc.
- Convert between fractions and decimals e.g. know which is more 65% or 30/60 and use this knowledge in problem solving contexts.

Ratio and Proportion

As in Year Four, children need to have an understanding of the term 'in every' and realise that it compares part to part. For example, 3 in every 4 cars are red means that of 4 cars, three are red, so ¾ of the total will be red. They also need to understand the term 'to every' which compares one part to the whole. So if 1 car to every 4 was red it would actually mean that 1 car in 5 or 1/5 would be red. If 2 cars to every 5 were red then it would mean that 2 in every 7 or 2/7 of the cars were red. The children will encounter this language in problems and need to be able to understand what it means and how it can be represented in fraction form. They will discuss how else this type of data can be expressed, using language such as 'twice as many as', 'half as many as...'. Once they are familiar with this language, they should begin to apply this knowledge to problems. To begin with, they should use practical equipment or draw diagrams to aid understanding.

For example, 'David has 3 apples for every 6 that John has' can be expressed as: David has 1/2 of the apples that John has; John has twice as many as David; David has 1/3 of the total; John has 2/3 of the total etc. Problems which allow discussion using this type of language will be given and eventually children will be expected to answer questions such as the following: 'What fraction of the larger shape is the smaller shape?'

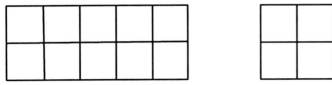

Figure 7.3

'A boy gets 1 star for every 4 stars his friend gets. If 20 stars are given in total, what fraction of the stars does the boy get?'

'In my cupboard, one tin in every 5 is dented. If there are 30 tins in the cupboard, how many are/aren't dented? '

In the pattern below, how many squares to crosses are there? (1 to every 2 or 1 in every 3).

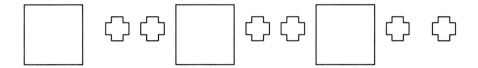

Figure 7.4

Your child will also need to understand how to scale up or down, establishing, for example, if there were 5 red sweets in every pack of 10, then there would be 10 red sweets in every pack of 20. Scaling involves increasing/decreasing a number or quantity by a given factor. Multiplication by 10 means scaling a number or quantity by a factor of 10 and making it 10 times as big. 'I have collected 25 stickers. My friend has 10 times as many. How many does my friend have?'

Negative Numbers

Your child will need to recognise negative numbers on a calculator display and should begin to devise their own number sequences which include negative numbers using the constant button. They should be able to count back beyond zero in steps of different sizes e.g. 5, 3, 1, negative 1, negative 3 etc and predict the next number in a sequence including negative numbers. They should also explain the rule or pattern in these sequences.

As in Year Four, they should order negative numbers and place them on a numberline and also fill in missing numbers on a numberline.[21] They should be able to state a negative number before or after another and state negative numbers between a given pair. Negative numbers will be used in the context of problems such as, 'The temperature in my freezer is minus 12. When I leave the door open, it rises by 3 degrees but then falls again by 5 degrees when I shut it again. What is the final temperature?'

Rapid Recall (Addition and Subtraction)

Although children will now require written methods more frequently, mental methods and knowledge of number facts are still integral to their work. Calculations strategies and number facts are needed to work quickly and efficiently. The following should be either known by heart or quickly worked out:

- Continue to know 2 digit pairs that total 100 (e.g. 34+66).
- Continue to know multiples of 50 that total 1000 (e.g. 350+650).
- Continue to gain new facts from known facts, e.g. 70+90=160; 700+900=1600; and be able to work out the inverse of these facts e.g. 160-90=70.

Children should also use known facts and understanding of place value to derive facts with decimal numbers:

- Know decimals (to one place) that total 1 (0.8+0.2).
- Know decimals (to one place) that total 10 (e.g. 1.4+8.6).

_____

[21] See page 89, Year Four.

- Derive quickly sums and differences, doubles and halves of 2 digit whole numbers and decimal numbers (e.g. work out 4.6+3.9 by relating it to 46+39 or 9.2-3.6 by solving 92-36; double 0.43 by doubling 43 etc).

Use known facts and understanding of place value to derive addition doubles:
- From double 1 to double 100.
- Double multiples of 10 from double 10 to double 1,000 (e.g. double 980 =1960). This basically means the double of any number up to 1,000 that can be said when counting in tens.
- Double multiples of 100 from double 100 to double 10,000 (e.g. double 6900=13800).
- Doubles of 2 digit decimals (e.g. 4.8+ 4.8).
- Know the corresponding halves for all of the above.

Mental Calculation Strategies (Addition and Subtraction)

These strategies are almost identical to those used in previous years. Each year group will use the same strategies with different numbers or in different contexts. If your child is finding it difficult to understand a particular strategy therefore, it can be useful to look back at previous years and try it in simpler contexts or with simpler calculations. Once your child grasps the idea, the numbers can be gradually changed to include larger (or, in the case of decimals, smaller) numbers.

- Start with the largest number when adding, unless another strategy is more appropriate.

- Add several small numbers, looking for pairs that make 10 or for doubles/near doubles and add these first. Also look for pairs that total 9 or 11 then adjust by 1. With addition of multiples of 10, look for pairs that total 100. (E.g. 40+20+60. Add 40+60 first as it is a number bond to 100 and then add on the 20; 30+50+40 – add 30+40 first as it is a near double which can be worked out without counting. This will total 70. Then add the remaining 50 to 70 to reach 120). Looking for patterns can also help with addition of several numbers. For example, 6+5+7+6 may be spotted as equivalent to 6x4.

- Use doubles and near doubles in mental addition, including decimals. E.g. 1.5+1.6= (double 1.5) +0.1=3.1.

- Partition (split) numbers, if appropriate, to make them easier to work with mentally.
  - Numbers may be partitioned according to place value (split the thousands, hundreds, tens, ones etc). E.g. 324+57 can be solved by partitioning the second number and counting on, i.e. 324+50+7:

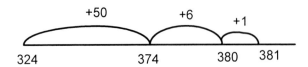

Figure 7.5

Alternatively, both numbers can be partitioned:
300+20+50+4+7=370+11=381.
  - The units may also be partitioned. In the above example, the 7 from 57 was split into 6 and 1 when added. Partitioning the units allows children to use the bridging strategy (see below).

- Bridge numbers through multiples of ten, a hundred or a thousand to make use of number bonds and known facts. For example, 628+55:

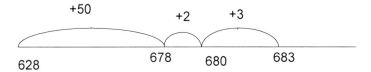

Figure 7.6

The 5 units have been partitioned in order to make use of the multiple of 10 (680) as a bridge (or stopping place). Now, instead of counting in ones from 678 to 683, children can use their number bonds and calculate mentally.

- Recognise that it is often easier to find the difference between two numbers by counting up from the smaller to the larger number instead of taking away the smaller from the larger.
  For example, 674-586; the 600 can be used as a bridge to count up from 586 to 674.

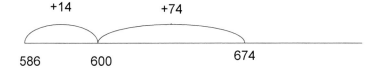

Figure 7.7

14 must be added to 586 to reach 600 then 74 more will reach 674 therefore the difference between 586 and 674 is 88 (14+74).[22]

- Use compensation. Continue to add/subtract 9, 19 etc by adding 10, 20 etc then subtracting 1. Now your child is becoming more confident with larger numbers, they should be able to round up a wider range of numbers, such as 274+96=370 (274 add 100 instead of 96 then subtract 4) or 4563-2997=1566 (4563 subtract 3000 then add 3).

- Be able to state three facts related to another e.g. 15.8+9.7=25.5 so 9.7+15.8=25.5 and 25.5-9.7=15.8 and 25.5-15.8=9.7. Children need to be very confident with inverse operations, gaining 4 facts from one initial fact. They should also be able to use this understanding of number to work out new facts e.g. 136+319=455 so 455-318=137 and 455-137=318.

- Recognise that when numbers similar in size are added, it may be quicker to multiply e.g. recognise that 28+30+26+32+34=30x5.

- Use knowledge of place value, number facts and rounding to estimate and check calculations. For example, use the inverse operation, perform an equivalent calculation, add in a different order or use knowledge of sums and products of odd and even numbers.

Although it is expected that children will solve a range of calculations mentally, as they practise their skills throughout the year they may need to record informally using

---

[22] For a more detailed explanation of subtraction as counting on (complementary addition) see page 50.

numberlines and jottings. These types of recording are not to be discouraged as they support the mental process and can clarify understanding.

Addition

Children should be able to work rapidly in their heads using a range of methods. By now children should have a good understanding of addition, so check that they:

- Know that, when adding whole, positive numbers, addition will make them bigger.
- Realise that addition can be done in any order e.g. 300+800=800+300, (300+800)+200 =300+(800+200).
- Understand that addition is the inverse of subtraction (i.e. it reverses it) and use it to check results.
- Respond rapidly to addition problems, explaining methods both orally and in writing.
- Approximate their answers before they start so that they have a feel for the size of the numbers with which they are working.
- Confidently add 4 digit numbers using an efficient written method.
- Record stages in calculations involving addition when working mentally.
- Confidently add decimal numbers (to 2 decimal places) using efficient written methods.

Subtraction

Children should have a good understanding of subtraction if they are to develop secure mental and written methods, so check that they understand that subtraction:
- Makes positive numbers smaller (except for subtraction of zero which leaves a number unchanged).
- Cannot be done in any order, unlike addition (i.e. that 5-3 is not equal to 3-5).
- Is the inverse of addition (i.e. it reverses it).
- Can be solved by taking away /counting back.
- Can be solved by finding the difference/complementary addition (counting up from the smaller to the larger number).

The Numeracy Framework for Year Five includes certain calculations for addition and subtraction. These calculations build on the work of previous years and, by the end of the year, should be tackled mentally, using the types of mental calculation strategies outlined in each year group. By now, children should be choosing the most appropriate from a range of strategies. Having said that, there may be several efficient methods to solve the same calculation and as long as it is quick and successful, it is acceptable.

1.Continue to add three 2 digit multiples of 10

E.g. 50+40+20=__

Often looking for doubles, near doubles or number bonds to 100 is the best strategy for this type of calculation. So in this example, 50+40 is a near double and equals 90 then 90+20=110.

2.Consolidate mental addition and subtraction of 3 digit multiples of 10

E.g. 570+250=__   620-380=__   610-__=240   __-370=240

Partitioning, making use of number bonds, doubles or near doubles and compensation may all be appropriate here. So, 570+250 could be seen as 570+200=770 then the 50 partitioned into 30 (to make 800) and 20 to reach 820. Another efficient strategy would be to say 500+250=750 then add 70. 620-380 can be worked out as 620-400+20 which is 240.

3. Add 3 or more 3 digit multiples of 100

E.g. 500+700+400=__   800+___+300=1500

Awareness of pattern in number and understanding of place value is required here. In the calculation 8+__+3=15 the solution would be 4, so therefore for 800+__+300=1500 the solution would be 400. Some children will be confident enough to add the hundreds, partitioning as appropriate. So 800+___+300=1500 may be solved by saying 800+200+100=1100 then subtracting 1100 from 1500 to find the answer.

4. Add or subtract a multiple of 100 (up to 900) to or from 3 or 4 digit numbers, crossing thousand boundary

E.g. 683+500=__   1263-400=__   300+__=1345

Mentally partitioning the numbers would be a good idea for these. Then use can be made of patterns in number so 1263-400, 12-4=8 so 1200-400=800 so the solution is 863.

5. Add or subtract 3 digit multiples of 10 to or from a 3 digit number, without crossing 100 boundary

E.g. 230+364=__   775- __=530

Mentally partitioning the number would be appropriate, 200+300=500; 30+64=94 so 594 is the answer. For subtraction, counting up is often the easiest strategy when working mentally so to work out 775-__=530, you could count up 70 from 530 to reach 600, then add 175 to reach 775 so the total added is 245 (70+175).

6. Continue to find what to add to a 3 digit number to make the next multiple of 100

E.g. 567+___=600

This consolidates the work carried out in Year Four. Counting up mentally should be done, perhaps by visualising the process previously practised on a numberline. In the above example, 3 would be added to make 570 then 30 to reach 600 so the total added would be 33.

7. Find what to add to a decimal with units and tenths to make the next whole number

E.g. 4.8+__=5.0

Use of pattern in number and understanding of place value is needed here. 48+2=50 so 4.8+0.2=5.0.

8. Find the difference between a pair of numbers either side of a multiple of 1000

E.g. 7003-6892=__
     8004-___=19

Counting up from the small number and using the multiple of 1000 as a bridge is the most efficient strategy here. So, if 108 is added to 6892 you reach 7000 then 3 more would reach 7003. In total 111 was added so that is the difference between the two numbers. Some children would record this on a numberline to support the mental process.

9. Add or subtract a pair of decimal fractions (with units and tenths or tenths and hundredths) including crossing units and tenths boundary

E.g. 5.7+2.5=__   0.56+0.72=__   0.63-0.48=__

Understanding of place value and using related number facts are required for these examples. 5.7+2.5 can be related to 57+25 to solve the first calculation as long as place value is understood.

Written Methods

A general introduction to written methods is given in Year Three[23] which explains the reasons for the changes to the teaching of written methods and the significance of the expanded methods now taught.

It is expected that most children will now be confident with an efficient written method and some children will be using standard compact methods. However, even if your child has been using the compact standard method, larger numbers are used in Year Five and it cannot be assumed that because a child can use a certain method to record 2 and 3 digit numbers that they will necessarily be able to use the same methods confidently with larger numbers. When introducing larger numbers therefore, it can be useful to revert to the expanded method to clarify understanding initially. Once their confidence grows, they can then transfer their knowledge of more compact methods to larger numbers. In Year Five children will be expected to use written methods of recording for addition and subtraction with numbers up to 10000 and also with decimals to two decimal places.

Developing Written Methods (Addition)

Partitioning

Two expanded methods for partition which were introduced and discussed in previous years are shown used with larger numbers below. They are useful to clarify understanding when still gaining confidence with larger numbers. However, the first example, in particular, is not particularly efficient with larger numbers and, if an expanded method is needed, the latter would be more appropriate for most children.

```
3 0 0 0 +  8 0 0 +   5 0 +   7
2 0 0 0 +  7 0 0 +   7 0 +   8  +
─────────────────────────────
5 0 0 0 +1 5 0 0 +1 2 0 +  15   =    6  6  3  5
─────────────────────────────
```

Figure 7.8

```
3   8   5   7
2   7   7   8   +
───────────────
5   0   0   0   (add thousands)
1   5   0   0   (add hundreds)
    1   2   0   (add tens)
        1   5   (add ones)
───────────────
6   6   3   5   (TOTAL)
───────────────
```

Figure 7.9

---

[23] Page 73.

<u>Compensation</u>.

This was presented as a written method in the original Numeracy Framework but, as discussed in previous chapters, has a higher profile as a mental rather than a written strategy.

```
     6  5  6  4
     9  4  8  0  +
   ─────────────
   1 6  5  6  4  (10000 added instead of 9480)
-        5  2  0  (subtract 520 as 9480 Is 520 less than 10000)
   ─────────────
   1 6  0  4  4  (TOTAL)
   ─────────────
```

Figure 7.10

<u>Compact Written Methods</u>

Once they are very confident with expanded written methods and have a secure understanding of larger numbers, children may be ready to use the compact method, which is the one that most adults will be familiar with from their own school days. However, as your child is working, it is important to continue to refer to tens as tens and hundreds as hundreds, just as in the expanded method. In the example below, when adding the tens it is important to say 60+80 rather than 6+8 so that your child retains an idea of the size of the numbers with which they are working.
E.g.

```
     5  6  4
     4  8  8  +
   ───────────
   1 0  5  2
     1  1
```

Figure 7.11

Children will be expected to use these methods with several numbers, often with differing numbers of digits and also with decimals. It is important that, when using any of these methods, children realise the importance of lining the digits up in the correct place. For example, if adding 5.8 metres to 240cm, the answer could only be correctly found if children knew that the 240cm was equivalent to 2.4m and recorded it in the correct place as shown in the example below:

```
   5 . 8  0
   2 · 4  0  +
   ──────────
```

Figure 7.12

This is something to pay particular attention to when working with decimals. The layout required for expanded methods of recording should have already given children the opportunity to discuss and understand this point.

Even if compact methods have already been introduced, expanded methods are usually returned to when introducing new concepts or when working with larger numbers. This is because, although the compact method is probably the quickest, it does not give a clear indication of the size of the numbers involved or link closely to mental methods. It is therefore not the best method for helping children to develop their understanding. It should be used once understanding is secure.

Developing Written Methods (Subtraction)

Children will work with both 3 and 4 digit numbers and will use the strategies as in previous years but with generally larger numbers.

Counting Up

```
2   6   7   2
1   3   8   5   -
        1   5    (to reach 1400)
    6   0   0    (to reach 2000)
    6   0   0    (to reach 2600)
        7   2    (to reach 2672)
1   2   8   7    (total counted on)
```

Figure 7.13

This strategy was introduced in previous years and may still be preferred by children who find subtraction difficult. It can be illustrated visually using a numberline (see below) in order to make the link between this and earlier methods explicit. If your child is using this method, then encourage them to make it as efficient as possible by reducing the number of steps taken and solving as much mentally as possible. In the above example, the calculation could be reduced further by counting up from 2000 to 2672 in a single step.

Figure 7.14

Compensation

An explanation of compensation is contained in previous chapters.

```
6 7 2
3 8 5  -
2 7 2   (400 subtracted instead of 385)
  1 5   (add 15 because 385 is 15 less than 400)
2 8 7
```

Figure 7.15

Figure 7.16

124

Exactly the same methods will be employed with 4 digit numbers.

Partitioning (Decomposition)

Decomposition is discussed in more detail in the relevant sections in previous chapters. By Year Five, it is hoped that some children will be able to reduce the number of steps in a calculation by partitioning and exchanging numbers mentally. Instead of recording the following:

```
                                  160
                          500      60    12
  6 7 2      6 0 0 + 7 0 + 2     600 +   70 +  2      5 0 0 + 1 6 0 + 1 2
  3 8 5 -    3 0 0 + 8 0 + 5 -   3 0 0 + 8 0 +  5 -   3 0 0 +  8 0 +  5 -
  _____    _____   _____   _____
                                                      2 0 0 + 8 0 + 7 = 2 8 7
```

Figure 7.17

Children would simply record:

```
  5 0 0 + 1 6 0 + 1 2
  3 0 0 +  8 0 +  5 -
  _____
  2 0 0 +  8 0 + 7  = 2 8 7
```

Figure 7.18

Standard Compact Method

Once children are able to partition and exchange mentally they may be ready to move onto the standard compact method shown below.

```
  5    16
  6    7  12
  6    7   2
  3    8   5  -
  _____
  2    8   7
```

Figure 7.19

As with addition, children need to be able to work with numbers with varying numbers of digits and with mixed units. For example, they must record and work out problems that have numbers recorded as both metres and centimetres or pounds and pence (e.g. £4.89-230 pence). It is important to know how to record these amounts in the vertical format so that the tens are below the tens, the hundreds below the hundreds etc. When 'borrowing' tens or hundreds in the compact method, it is important to refer to them as the actual numbers so that your child retains an idea of the size of the numbers involved. In the example above, you would say that we move 100 from the 600 to make 160 in the tens rather than moving one from the 6 to give 16 in the tens. Children are far more likely to spot errors that occur when they work in this way.

Multiplication

It is important to understand:

- When multiplying positive whole numbers, multiplication makes a number larger.
- Multiplying by one leaves a number unchanged.

- Multiplying by zero=0.
- Multiplication as repeated addition e.g. 72x3=72+72+72 or 72 sets of 3.
- Multiplication can be done in any order and the answer will remain the same. E.g. 45x7=7x45; 16x12=(2x8)x12=2x(8x12).
- Partitioning can be useful to help solve multiplication problems as well as addition and subtraction. Partitioning numbers and multiplying each part will reach the same answer as multiplying the whole number, e.g. 34x6=(30x6)+(4x6).
- Division is the inverse of multiplication and can be used to check their work.

Children need to be able to count forwards and backwards from any number in whole number and decimal steps. For example, count on in steps of 0.1 to 5.0 and back. To make this more challenging, children should also be asked to start at different starting points, rather than zero, and see if they can still count in the steps given. When counting backwards, the steps should extend beyond zero and into negative numbers. As well as counting on in this way, children should also be able to extend and continue given sequences, explaining the rules and finding missing numbers. In schools children may be asked to colour patterns on different size grids in order to see the variations and spot patterns. Counting on in steps such as 19, 25 etc should also be tried. By now, children should be able to find a quick way to do this (e.g. add 19 by adding 20-1 or add 25 by adding 20 then 5). Using the constant button on the calculator is another way to investigate multiples and patterns in number.

It is important to be able to understand the significance of multiplying and dividing by 10,100 and 1000 (i.e. that the digits move one, two or three places to the left or right). General rules can be learnt which help to improve understanding e.g. to multiply by 100, simply multiply by ten then ten again.

Children need to be able to relate their understanding to fractions when working with tenths and hundredths so they should be aware that 1/10 = 0.1 and 2/10=0.2 etc.
Rules regarding odd and even numbers and multiples of numbers (up to 99) should be explored. For example, children should know the rules of divisibility:
  o Multiples of 100 end in 00.
  o Multiples of 10 end in 0.
  o Multiples of 5 end in 5 or 0.
  o Multiples of 2 end in 2, 4, 6, 8 or 0.
  o They should also recognise that if the last 3 digits of a number are divisible by 4 then it is a multiple of 4.

They should be able to find all the pairs of factors of a 2 digit number, find common multiples for 2 given numbers (e.g. 'What are the common multiples of 6 and 9?') and use their understanding of factors to simplify or check calculations. E.g. 16x12=16x(3x2x2).
Children in Year Five begin to learn about square numbers by relating them to drawings of squares. They should be able to tackle questions such as 'What is 8 squared?' They may also begin to understand and use brackets in their work, realising that brackets indicate the part of the calculation that is to be done first. For example, 4+(7x3) =25 but (4+7) x3=33. When multiplying, they need to be able to work rapidly, using mental strategies and, if necessary, jottings to multiply 2 digit numbers by single digit numbers.

Division

Children need to understand that:
- When working with positive whole numbers, division makes them smaller.
- Division cannot be done both ways (e.g. 35 divided by 5 is not the same as 5 divided by 35).
- A number cannot be divided by 0.
- Division can be seen as sharing equally or grouping (repeated subtraction).
- Sharing may often be appropriate for working with smaller numbers but grouping is better for working with larger numbers.

- Multiplication is the inverse of division and use this to check their work.
- Fractions and division are related e.g. understand that 1/3 of 24 =24 divided by 3 or 24/3.
- Children should be able to use jottings and mental strategies to divide a 2 or 3 digit number by a single digit number.
- They should begin to give a quotient (answer) as a fraction (e.g. 43 divided by 9=4 and 7/9) or a decimal fraction when dividing by 2, 3, 4, 5 or 10 (e.g. 351 divided by 10=35.1) and when dividing pounds and pence by a small whole number (e.g. £5.40 divided by 4=£1.35).
- Children should be able to interpret the quotient on a calculator e.g. 8.4 as 8.40 in money and round other decimals to the nearest whole number, so, for example, know that 9.714 is between 9 and 10, but nearer to ten.
- It is also important that children learn to use division in problems and can decide whether to round up or down depending on the context of the problem. For example, 'A restaurant buys 427 plates. Only 6 plates fit into a box. How many boxes will the restaurant owner need in order to carry all the plates?' This problem requires rounding up as all the plates must be carried and therefore an extra box is needed for the remainder of the plates. Although the answer to 427 divided by 6 is 71 remainder 1, the answer to this calculation must be rounded up to 72. In the next problem, however, the answer must be rounded down. 'A shopkeeper sells packets of sweets with 6 sweets in each packet. If he has 427 sweets altogether, how many full packets of sweets can he make?' The answer to the calculation is still 71 remainder 1 but this time the answer has to be rounded down to 71 because the remainder will not count as a full packet.

Rapid Recall (Multiplication and Division)

Again, the instant recall of facts cannot be over-emphasised. It is important to:
- Know all tables facts up to 10x10 (including x0 and x1) and corresponding division facts.
- Know the squares of all numbers from 1x1 to 10x10.
- Know that halving is the inverse of doubling.
- Know doubles of all numbers up to 100.
- Know doubles of all multiples of 10 up to 1000.
- Know doubles of all multiples of 100 up to 10,000.
- Know corresponding halves for all of the above.
- Be aware of the effect of multiplying or dividing whole numbers and decimals by 10, 100 or 1000.

Mental Calculation Strategies (Multiplication and Division)

- Use knowledge of place value to derive new facts E.g. 4x5=20 so 4000x5=20,000.
- Partition numbers e.g. double 67= double 60 then double 7, 120+14=134.
- Partition to multiply a 2 digit number by a single digit number, e.g. 45x7= (40x7)+(5x7).
- Be aware of the relationship between numbers when they are multiplied e.g. realise that 12x5=6x10 or 6x10=3x20 i.e. in a given multiplication the same product can be reached by halving one factor and doubling the other. Your child should be able to solve such calculations by doubling a factor ending in 5 and halving the other.
- Be able to gain new facts by doubling e.g. double the 2 times table for the 4 times table; double the 4 times table for the 8 times table; double the 8 times table for the 16 times table etc.
- Use combinations of multiplication to work out calculations
  - E.g. use knowledge of doubles to find facts for the 25 times table then work out other facts by combining these. For example,25x25= (10x25)+(10x25)+(5x25) or 25x25= (16x25)+(8x25)+(1x25).
  - Work out 12 times table by adding 10 times and 2 times tables.

- Halve an even number to multiply it then double the answer, e.g. 46x3; 23x3=69 and double 69 is 138 so 46x3=138.
- Know that to multiply by 50, multiply by 100 then halve.
- Work out that to multiply by 25, multiply by 100 then quarter.
- Gain the facts for the 16 times tables by doubling those of the 8 times table.
- Find sixths by halving thirds, twentieths by halving tenths etc.
- Use understanding of factors to simplify or check calculations,
  e.g. 15x6= 15x3x2.
     15x3=45; 45x2=90 so 15x6=90.
  The same strategy can be used for division:
  90 divided by 6 =90 divided by 3 divided by 2.
  90 divided by 3=30; 30 divided by 2=15.
- Use compensation e.g. to multiply by 19 or by 21 multiply by 20 then add or subtract the number.
- Use knowledge of the inverse, realise that knowing one fact also means you can quickly derive other related facts. So 23x3=69; therefore 3x23=69; 69 divided by 23=3 and 69 divided by 3=23; 12x6=72 so 1/6 of 72=12 and 1/12 of 72 =6.

The following calculations for multiplication and division are examples of what children in Year Five should be able to solve mentally by the end of the year.

1.Multiply a 2 digit multiple of ten by a 3 digit multiple of 100

E.g. 30x400   40x700

Use known facts and understanding of place value to work this out. E.g. 3x4=12 so 30x400=12000.

2.Divide a 4 digit multiple of 100 by 1000, 100 or 10

E.g. 8200 divided by 100 or   __divided by 100=8 etc.

Understanding of place value and patterns in number are needed for this.

3.Double any multiple of 5 up to 500

E.g. Double 435; Double 430=860 and double 5=10 so 860+10=870.

4.Halve any 3 digit multiple of 10

E.g. 150x ½; Halve 100 then halve 50; 50+25=75.

5.Multiply a 2 digit multiple of 10 or a 3 digit multiple of 100 by a single digit number

E.g. 400x9   60x8

Again, use known facts to solve this. E.g. 4x9=36 so 400x9=3600.

6.Multiply a 2 digit whole number by any single digit number (crossing tens boundary)

E.g. 24x3   17x4

These are generally best solved by partitioning. E.g. 24x3; 20x3=60 and 4x3=12; 60+12=72. Sometimes compensation can be used to solve these. E.g. 59x5; 60x5=300 then subtract 5 because 59 is one less than 60. Using other facts then doubling or halving can also be appropriate when multiplying by 4, 8 or 5 (see 'Mental Calculation Strategies').

Developing Written Methods (Multiplication)

As in all calculations in Year Five, children are encouraged to approximate their answer first then work it out. By the end of Year Four it is expected that children will be confident with the expanded method for multiplication when multiplying a 2 digit by a 3 digit number. In Year Five they should gain confidence using more compact methods to multiply and divide with smaller numbers. However, when larger numbers are introduced, the expanded method is often used initially to ensure that children have an understanding of the numbers involved.

Grid Method

The expanded method shown below links closely to the mental methods taught and can be carried out when the numbers involved are too large to work with mentally or if the child is unsure of the compact method. It is similar in many ways to expanded methods of recording addition and subtraction in that it involves partitioning the number to carry out the operation then recombining it at the end. This method is known as the grid method.

Short Multiplication

458x8

x	400	50	8
8	3200	400	64

Figure 7.20

3200+400+64=3664.

The grid can also be set out with the larger number down the side which can make it easier to add the products and is the way now suggested by the revised Numeracy Framework.

x	8
400	3 2 0 0
5 0	4 0 0
8	6 4
(Total)	3 6 6 4

Figure 7.21

Total= 3200+400+64=3664

Once the grid method is understood, the grid can be changed to become more like a vertical calculation as shown:

```
x │ 4 0 0 + 5 0 + 8
──────────────────────
              8
          3 2 0 0  (400x8)
            4 0 0  (50x8)
              6 4  (8x8)
          3 6 6 4  (Total)
```

Figure 7.22

In Year Five children should be able to multiply simple decimals with one decimal place by partitioning them.

E.g. 6.7x4   6.0x4=24.0
             0.7x4=_2.8
             Total=26.8

It can be set out in the grid if preferred.

```
x       │      4
─────────────────────
6       │  2   4
0 .7    │      2 . 8
Total      2   6 . 8
```

Figure 7.23

Expanded Method

Expanded short multiplication follows on quite naturally from this and can be recorded in one of the following ways:

```
400+ 50 + 8                 4 5 8
          8 x                   8 x
─────────────────           ─────────────────
    3 2 0 0 (400x8)          3 2 0 0 (400x8)
      4 0 0 (50x8)             4 0 0 (50x8)
        6 4 (8x8)               6 4 (8x8)
    3 6 6 4 (Total)          3 6 6 4 (Total)
```

Figure 7.24

The Standard Compact Method

It is expected that many children would be confident with expanded methods for short multiplication (for multiplying a 2 digit number by a single digit number) by the end of Year Four and in Year Five should gain an understanding of compact short multiplication (shown below). At first the compact method would be introduced when multiplying single digits by 2 digit numbers and expanded methods would be used for larger numbers. Once the children develop their confidence they may also be able to use the compact method for short multiplication with 3 digit numbers.

The compact method should be encouraged as soon as children have a secure grasp and understanding of expanded methods and can use them confidently with a range of numbers.

It is important that when using it, however, that children still retain an idea of the size of the numbers so when multiplying the tens and linking it to the known fact 5x8=40, make sure that you state it is 50x8=400.

```
      4   5   8
              8   x
  ─────────────────
  3   6   6   4
  ─────────────────
      4   6
```

Figure 7.25

If children begin to make errors when using the compact method, they should return to the expanded method of recording.

Long Multiplication

Grid Method

The expanded method for long multiplication (2 digit number x 2 digit number) is taught initially using the grid method. Again, children are expected to approximate their answer first.

73x45. Estimate: 70x50=3500.

x	40	5	
7 0	2800	350	3150
3	120	15	135
		Total	3285

Figure 7.26

The revised framework suggests that the larger number in the calculation is placed in the left hand column to make the addition of the partial products easier. This layout can easily be adapted to the layout below, also advised in the revised Numeracy Framework, as a 'stepping stone' leading to the vertical expanded method.

	7 0	3	
x	4 0	5	
	2 8 0 0	3 5 0	3150
	1 2 0	1 5	135
		(Total)	3285

Figure 7.27

Expanded Method

Once children are confident with long multiplication using the grid methods, they can move on to using the following, vertical expanded method.

```
    7 3
    4 5 x
2 8 0 0 (70x40)
  1 2 0 (3x40)
    3 5 0 (70x5)
      1 5 (3x5)
3 2 8 5 (Total)
    1
```

Figure 7.28

Because all the numbers are added at the end rather than adding them in stages, this expanded method involves less recording (bear in mind that the calculations at the side of each number do not *need* to be recorded by the children).

<u>The Standard Compact Method</u>

The more compact method, which (the revised framework states) children are now to aim for by the end of Year Five, is shown below:

```
    5 6
    2 7 x
1 1 2 0 (56x20)
  3 9 2 (56x7)
1 5 1 2 (Total)
    1
```

Figure 7.29

It is worth mentioning that the carry digits in the partial products of 56x20= 1120 and 56×7=39 are usually carried mentally.

As with all compact methods, if children are not working confidently it is preferable to revert to expanded methods, which, although more time consuming and therefore less efficient, are a means of gaining accurate answers in a way that deepens understanding of the processes and numbers being used.

<u>Developing Written Methods (Division)</u>

By the end of Year Four or the beginning of Year Five, it is expected that children will be starting to use more compact forms of recording for short division of 2 digit numbers by a one digit number. As with multiplication, children may revert to using more expanded methods (see Year Four) if they are making errors with the more compact forms of recording. It is important that they do this in order to consolidate their understanding. They may also revert to expanded methods when using larger numbers or when more difficult calculations are initially introduced (demonstrated later in this chapter). To use the more compact form, your child will need to be confident with their times tables and the corresponding division facts. They will also need to be able to subtract multiples of ten mentally and have a secure understanding of place value and partitioning.

Short Division: Partitioning

96 divided by 4. To find how many times 4 will divide into 96, children must be encouraged to find the highest multiple of ten (below 96) that is also a multiple of 4. We know that 10x4=40 so 20x4 must equal 80. This is the highest multiple of ten which is also divisible by 4 (and below 96). 96 is therefore partitioned into 80+16.

$$4 \overline{)\ 80 + 16} \quad \begin{array}{c} 20 + \ 4 \end{array}$$

Figure 7.30

This can gradually be shortened to:

$$4 \overline{)\ 9\,{}^1 6} \quad \begin{array}{c} 2\ \ 4 \end{array}$$

Figure 7.31

As they record in this way, children are asked 'How many fours divide into 90 so that the answer is a multiple of ten?' The answer is 20 fours or 80 and there are 10 remaining. (To work this out children are encouraged to use their tables, 10x4=40 so 20x4=80). The remaining ten are added to the 6 to give 16 then 'How many fours are in 16?' is asked. This gives the answer 4.

Should your child find this method difficult to understand then it would be wise to practise dividing 2 digit numbers by single digits using one of the more expanded methods (see Year Four).

Using Multiples of the Divisor

When children begin to work with larger numbers, expanded methods are generally returned to in order to ensure understanding so when division of 3 digit numbers by a single digit are introduced this method (sometimes known as 'chunking', as chunks are subtracted) may be employed. It involves repeated subtraction, or grouping, using knowledge of tables facts to subtract larger multiples. Multiples of ten are still used as a starting point.

For example, 567 divided by 8.

567 divided by 8

```
      5   6   7
  -       8   0   (10x8)
      ─────────
      4   8   7
  -   1   6   0   (20x8)
      ─────────
      3   2   7
  -   1   6   0   (20x8)
      ─────────
      1   6   7
  -   1   6   0   (20x8)
      ─────────
              7
```

Figure 7.32

The question to be asked in order to help your child decide on the number to be subtracted is 'Which multiples of ten (below 567) can you find which the divisor (in this example, 8) will divide into?' These multiples are then subtracted. Answer: 70 remainder 7, since it was possible to obtain 70 groups of 8 from 567 and there were 7 remaining.

Initially it may be quite difficult for children to find an approximate answer for this type of calculation as rounding the numbers (for example, to say 600 divided by 10) does not always lead to a very accurate estimate. It is worth encouraging your child to approximate an answer for this type of calculation by finding multiples of ten which will divide by 8. 560 divided by 8=70 and 640 divided by 8=80 so the answer must lie between 70 and 80. Once children become confident with approximating in this way, they will have a useful starting point for subtracting. If they realise that 70x8=560, they can begin by subtracting 560 which is far quicker.

To use this method of division it is essential to have a good understanding of division as grouping, which is actually the same as repeated subtraction (just as multiplication is actually the same as repeated addition). If your child finds the concept of repeated subtraction difficult then problems requiring division as grouping should be attempted, using smaller numbers. Demonstrating division on a numberline to demonstrate the groups can be helpful and discussion of the groups being taken away should emphasise the concept of repeated subtraction. For example, in the following problem a numberline can be drawn or the numbers represented pictorially.

'Jason has 93 football stickers and each page of his football book can hold 7 stickers. How many pages of his book can he fill with his 93 stickers?'

The calculation is 93 divided by 7, or how many sevens can he get from 93? It can be illustrated as follows:

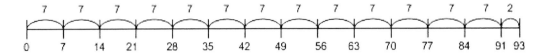

Figure 7.33

It is necessary to repeatedly subtract 7 from 93 and the easiest way to do this is to count in steps of 7 until you reach 91 then recognise that there are 2 left over. However, the illustration above is very laborious and time consuming so it is easier to use tables facts and represent it as in one of the examples below:

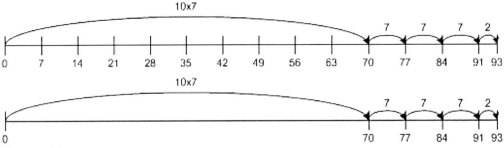

Figure 7.34

It can be pointed out that because your child (presumably) already knows that 10x7=70, they need only count on from 70 to 93 in sevens, instead of counting in sevens to reach 70.

This makes a link to the non-standard method by subtracting ten sevens at once instead of repeatedly subtracting 7. Using the numberline to illustrate division is important as it helps children to understand the repeated subtraction and to see the link between multiplication and division. Once your child is confident with the concept of grouping when dividing, they should grasp the fact that subtracting multiples of the divisor is far quicker than repeatedly subtracting the divisor.

The expanded written method below is very similar to the previous method but is set out as a division instead of vertically as a subtraction. 453 divided by 6

```
6 ) 4  5  3
  -  3  6  0   (60x6)
        9  3
     -  6  0    (10x6)
        3  3
      -  3  0   (5x6)
           3
```

Figure 7.35

Answer is 75 remainder 3.

As children gain confidence with this method they become more adept at working flexibly to find the highest multiple of a number possible. They may begin by subtracting multiples of ten but then double and treble these to make the multiples larger and therefore make the calculation quicker, so, for example, initially they may say 6x6=36 but use their knowledge of pattern in number to move onto saying 60x6=360. It is important for children to realise that chunking is inefficient if there are too many subtractions and therefore the highest multiples possible should be subtracted at each stage. The approximation made before the start of the calculation should help to give children an idea of the size of the number to be subtracted, thus making the calculation more efficient.

Partitioning (Short Division)

This was included earlier in the chapter to divide a 2 digit number by a single digit. Once larger numbers have been explored using the expanded method above, it can be reintroduced to divide a 3 digit number by a single digit.
For example, 455 divided by 6.

Approximate: 70x6=420 and 80x6=480 so the answer must lie between 70 and 80.
Because we already know that 70x6 is 420, we can partition 455 into 420+35.

```
     70 +  5  R5    = 75R5
  6 )420 + 35
```

Figure 7.36

The calculation is then solved by asking 'How many sixes will divide into 420?' (We know it is 70). Then 'How many sixes will divide into 35?' (5 with 5 remaining). So the answer (quotient) is 75 remainder 5.

To use this method, children need to have strong mental skills; their knowledge of times tables facts, partitioning and place value must be extremely secure. Once children are using this method confidently, they can begin to work more efficiently by using the standard compact method for short division shown below. As your child is working, remember to refer to the actual numbers, e.g. 'How many times will 6 divide into 450?' rather than 'How many times will 6 divide into 45?' This helps children to retain an understanding of the size of the numbers.

$$6 \overline{\smash{)}45\ {}^3 5} = 7\ 5\ R5$$

Figure 7.37

In summary, although the breadth of the curriculum increases in Year Five and more in depth work on topics such as percentages and decimals is introduced, the basic mental strategies are still used regularly as part of mental calculations. They do therefore require consolidation and practice. This can often be done through the application of skills and facts during problem solving activities as well as through rote learning. Discussion should still be a high profile part of learning but, rather than explaining methods and alternative strategies, many children will have moved beyond this and will be talking about ways to apply what they know logically in problem solving situations. Many children will now be using compact written methods for addition, subtraction, multiplication and short division. These standard methods, however, still rely on instant recall of many facts. It is therefore useful for children to continue to practise these both at home and at school so that they are not forgotten.

Chapter 8 : Year 6

Much of the Year Six curriculum involves consolidating and applying strategies from previous years. They are either used with larger numbers (or smaller decimals) or applied in new situations. To avoid large sections of repetition, therefore, unless a strategy or concept is completely new, there are only brief explanations of activities to help in this chapter. Instead I have focused primarily upon what your child will be covering and is expected to know in Year Six. Detailed explanations of the activities suggested or referred to in this chapter can, however, be found in the previous two chapters. So, for example, if you want to help your child with place value, then read the relevant sections in the previous two chapters to ensure they have the understanding required, then look at what is required in Year Six, using the same types of ideas and equipment as in previous years. This chapter also differs from other chapters in that it includes, not only what is expected from the average child, but also what is expected for the higher achievers who would be likely to attain a level 5 in the end of year SATS. This more challenging material is included to stretch your child if you feel they are ready. However, it should only be attempted when children are secure in their understanding of the Year Six curriculum.

Place Value

It is hoped that the work based around partitioning[24], positioning, ordering and rounding numbers carried out in previous years will have provided children in Year Six with a secure grasp of place value. This understanding is fundamental to their work both with whole numbers (integers) and with decimal numbers. In Year Six they will be expected to:

- Read, write and place in order any whole number including 2 digit negative numbers and decimals to 3 places.
- Find the difference between positive and negative numbers or two negative numbers in the context of a problem.
- Round any whole number to the nearest 10, 100 or 1000 and be able to say which would be the most sensible as an approximation in different contexts. For example, 'Would you round the size of the crowd at a concert to the nearest 10, 100 or 1000?'
- Use known facts along with their understanding of place value to solve calculations involving decimal numbers. For example, work out that because 54 divided by 9= 6 then 5.4 divided by 9 will be equal to 0.6.

If you find that your child struggles with any of the above, check that they can carry out the activities suggested in Years Four and Five with 4 and 5 digit numbers before carrying out similar activities (such as those suggested with digit cards, place value arrows and numberlines) in order to identify and address areas of difficulty.

Decimals

Assuming that your child can confidently work with decimals up to 2 places and has no difficulty with the activities suggested in Years Four and Five,[25] they should now be ready to extend their understanding by working with smaller decimals. The same types of activities can be repeated with decimals with up to three decimal places. You should help your child to:

- Use decimal notation for tenths, hundredths and thousandths.
- Record decimal fractions to 3 decimal places using decimal notation, e.g. 'Write the decimal fraction equivalent to 3 tenths, 4 hundredths and 6 thousandths.'

---

[24] See glossary.
[25] See pages 89 and 114.

- Partition and recombine decimal fractions (to three decimal places) and say what each digit represents. For example, know that in 8.675 the 8 represents 8 units, the 6 represents 6 tenths, the 7 is 7 hundredths and the 5 represents 5 thousandths.
- Order decimal fractions to three places from either the largest to smallest or smallest to largest (e.g. 15.789, 36.345, 15.234, 11.725 etc) and explain their thinking.
- Position decimal fractions to three places on numberlines.
- Give a decimal fraction between 2 others e.g. 'Give a decimal fraction between 5.75 and 5.76.'
- Round decimals to the nearest whole number or tenth (i.e. one decimal place).
- Continue or fill in the missing parts of number sequences involving decimals. For example, continue the pattern: 2.92, 2.94, 2.96, 2.98 etc. Explain the pattern or rule.
- Use a calculator to change one decimal fraction to another in one step. For example, know what to add to change 3.257 to 3.857; know what to subtract to change 6.469 to 6.461; know what to multiply 0.67 by in order to reach 6.7 etc.
- Convert between units when using decimals, for example, when working with money or measures convert 400ml to 0.4l; 6cm to 0.06m;250g to 0.25kg.
- Write decimals as fractions e.g. 0.004= 4/1000 or 6.567=6 and 567/1000 and fractions as decimals e.g. 37/100= 0.37 or 4 and 78/100 = 4.78.
- Interpret fractions as decimals when entered into a calculator, predicting the decimal first. For example, 1/3 =0.3333;1/6=0.666;1/8=0.125 etc.
- Use a calculator to compare fractions (displayed as decimals) and say which is larger or smaller e.g. which is smaller 4/5 or 14/17?

If you do feel that your child needs support with the understanding of decimals, remember to give them the opportunity to revert back to the use of practical activities and equipment as they work. Make sure that their understanding of the place value of decimal numbers is secure by using reverse place value arrows and linking examples to familiar contexts such as money (ten pence and one penny coins can be useful for representing tenths and hundredths of a pound). Using a numberline such as the one shown below (figure 8.1) can help to show the relationship between tenths and hundredths. For example, it can be used to show how 20 hundredths (0.20) is equivalent to 2 tenths (0.2). By labelling the same numberline differently (beginning at 0 and ending with 0.1, for example) it could also be used to show thousandths.

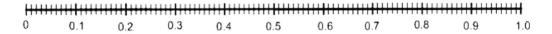

Figure 8.1

Continue to present problems in familiar contexts. For many children (and adults!), maths becomes difficult because it is too abstract. However, once it is linked to a context or a prop of some sort, it can be far more accessible.

<u>Fractions</u>

As with decimals, your child must be confident with what has been taught in Year Four and Five[26] as this will form the foundation for what they learn in Year Six. They should be able to:

- Count on in multiples of, for example, one third or one fifth, extending the sequence past one into mixed numbers.

---

[26] See pages 91 and 115.

- Identify the rule for given sequences and predict the next number, explaining why.
- Express a larger number as a fraction of a smaller one (e.g. 8/5) using practical equipment or diagrams. Understand the relationship between improper fractions and mixed fractions. For example, be able to state how many halves are in 1 ½ or 9 ½; how many thirds in 4 and 2/3; know that 18=1 ½ of 12 etc.
- Convert mixed fractions to improper fractions and vice versa. E.g. 38/6= 6 and 2/6 or 6 and 1/3.
- Order mixed fractions and proper fractions and give a number half way between two fractions.
- Use simple relationships between fractions e.g. ½ is double ¼ and 3x 1/6; 1/3 is twice as much as 1/6 etc.
- Continue to relate fractions to division and multiplication, including giving answers as fractions e.g. 18 divided by 4= 18/4 or 4½; know that 38 divide by 6 =38/6; realise that 7/2 is the same as 7x ½. Apply this knowledge in problem solving situations.
- Give fractions of numbers and quantities e.g. '3/10 of 80?' '7/10 of 200?' '5/6 of 36?' etc. 'What fraction of 1kg is 400g?'
- Explore and explain patterns of equivalent fractions, e.g. ½=2/4, 3/6, 4/8 etc 1/3=2/6, 3/9 etc.
- Know how to simplify fractions by cancelling common factors e.g. reduce 6/24 to ¼ by dividing both the numerator (top number) and the denominator (bottom number) by the same number and use this to order and compare. They should also realise that by multiplying both the numerator and the denominator by the same number an equivalent fraction can be made, e.g. 1/5 can be made into 5/25 by multiplying both by 5.
- Compare or order simple fractions by converting fractions to a common denominator, so give a fraction between ¼ and 1/3 by multiplying both to gain a common denominator. (E.g. multiply 1/4 by 6 to gain 6/24 and multiply 1/3 by 8 to gain 8/24. A fraction between the two is therefore 7/24).
- Relate fractions to work on ratio e.g. 1 in every 6 =1/6.
- Solve problems involving fractions, using a calculator where appropriate.
- Order a set of fractions by converting them to decimals on a calculator and rounding them if necessary. (This is now included in the section of the revised framework entitled 'Year 6 Progression to Year 7', which is aimed at higher achievers who are likely to be working within level 5 of the National Curriculum).

Percentages

Children need to consolidate and extend their understanding of the work covered in Year Five.[27] Then you can help your child by ensuring they can:

- Answer simple questions involving percentages. For example, state the percentage of girls in a class if the percentage of boys is given, i.e. know that if 35% are boys then 65% must be girls.
- Express one quantity as a percentage of another, e.g. express 600g as a percentage of 1kg.
- Represent simple fractions (e.g.1/2, 1/3, 2/3, 1/4, 3/4 etc) as percentages.
- Understand relationship between fractions, decimals and percentages. State percentages as fractions and decimals, (e.g. 35%=0.35 or 35/100) and vice versa.
- State that one tenth can be expressed as 0.1 or 10%, one hundredth is 0.01 or 1%.
- Understand that 1/3 is 33 1/3% and 2/3 is approximately 67%.
- State percentages that are greater or less than ½.
- Shade a percentage of a shape, using their knowledge of fractions to help them.

---

[27] See page 116.

- Calculate a percentage of an amount, for example, 16/80 by cancelling (dividing the 80 by 8 to reach 10 then dividing the 16 by 8 to reach 2) then knowing that 2/10 is the same as 20/100 or 20%. Mentally find an amount such as 40% of £6 (40% of £1 is 40p so 40px6=£2.40). Use a calculator to find other percentages. (This involves putting in the amounts involved in fraction form and interpreting the decimal answer as a percentage. It may involve rounding up or down to the nearest whole number).
- Use relationships between numbers e.g. know that 12.5% is half of 25% which is half of 50%.
- Solve problems which include decimals, fractions and percentages. E.g. 'A cake normally cost £5.00 but in the sale there was 40% off. How much did it cost?'
- Decide when it is appropriate to use a calculator to solve problems and know how to enter and interpret information accurately.
- Children working at level 5 of the National Curriculum will begin to use a calculator to work out a percentage increase or decrease in a problem solving context. They will also begin to recognise approximate proportions of a whole and describe them using fractions or percentages, for example, in the context of a pie chart.

Ratio and Proportion

As in Year Four and Five (see examples in previous chapters), children are expected to be able to understand and use the language associated with ratio and proportion and to use this understanding to solve problems. They need to understand:
- The comparison of part to part, e.g. two to every three is the same as saying two in every five.
- The comparison of part to the whole, e.g. one in every.

Many problems involving ratio and proportion are based around scaling up or down. For example, a recipe for 4 people may be given and children may be asked to scale up the amounts to make it suitable for 12 people. Similarly, the cost of 7 packets of sweets may be given and your child asked to work out the cost of ten packets. Children working within level 5 of the National Curriculum will begin to use ratio notation e.g. know that one teacher for every 3 children can be recorded as 1:3 and reduce a ratio to its simplest form. In problems they will be asked to divide quantities into two parts in a given ratio.

Negative Numbers

As in Year Four and Five (see relevant sections in previous chapters), children are expected to understand and use negative numbers in context and be able to order positive and negative numbers, placing them on a numberline correctly. They should also be able to find the difference between a positive and a negative number or 2 negative numbers in context.

Addition and Subtraction

All the knowledge required in previous years about the effects of addition and subtraction is required. Children are also expected to:
- Add and subtract 2 digit numbers mentally and solve additions and subtractions with some 3 and 4 digit numbers mentally if a particular strategy is appropriate. For example, use compensation (see mental calculation strategies below) to solve 4556-999.
- Mentally add 3 or more (2 digit) multiples of 10, and 3 or more 2 digit numbers.
- Add and subtract decimals to one place mentally.

Mental Calculation Strategies (Addition and Subtraction)

These strategies have been used in each year group but with different numbers and in different contexts. Children in Year Six would be expected to use and apply them in problem solving contexts. For children who are working within level 5 of the National Curriculum, mental methods are also expected to be employed when working with decimals, fractions and percentages.

- When adding by counting on, consider starting with the larger number, unless another strategy is more appropriate.

- When adding several numbers, look for pairs to make ten, one hundred or one thousand, doubles and near doubles and add these first, in order to speed up calculations. Also, look for pairs that total 9 or 11 in order to add 10 and adjust.

  E.g. 4+3+5+12+3 could be worked out more quickly by using doubles and a number bond to 9: 12+(3+3)=18. 5+4=9 so add 10 and subtract one as a quick way to add 9.

- When adding several numbers, also look for pairs that will total a multiple of 10. E.g. 36+19+24. Add 36 and 24 to total 60 then add on 19 to reach 79.

- Use knowledge of near doubles in a range of calculations, e.g. 421+387=808 can be seen as (double 400) +21-13. 1.5+1.6 can be seen as (double 1.5) + 0.1.

- Partition (split) numbers mentally

  o According to place value. This basically means splitting the number according to thousands, hundreds, tens and units. Once they are split, it is easier to add the thousands, then the hundreds etc mentally. E.g.
    6792+3201=6792+3000=9792.
              =9792+200=9992.
              =9992+1=9993.
  o Partition units in calculations in order to make use of the bridging strategy (see below). This is when units are split in order to make use of a multiple of 10, 100 or 1000 as a bridge (or stopping place). Partitioning the units in this way allows children to use number bonds to solve calculations instead of counting in ones. In figure 8.2, the calculation 628+55 is solved by partitioning the 55 into 50+5 and adding 50. The 5 is then partitioned into 2+3 so that, instead of counting on in ones from 678 to 683, the calculation can be solved mentally using knowledge of number bonds.

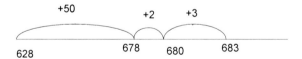

Figure 8.2

- Use the bridging strategy. Use multiples of 10, 100 or 1000 as bridges (or stopping places) in order to make mental calculations easier. In figure 8.3 the numbers 2800 and 3000 are used as bridges so that, instead of counting up from 2785 to 8000 in one step, three smaller steps can be used, thus making the calculation easier to solve mentally. This can be used in both addition and subtraction.[28]

---

[28] See 'Mental Calculation Strategies' sections in previous chapters for more detailed explanations of partitioning and bridging.

- Find the difference between 2 numbers by counting up through the next multiple of ten or hundred (again using 'bridging') from the smaller to larger number.[29]

8000-2785=

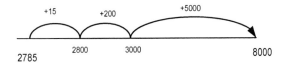

Figure 8.3

- Use compensation with a range of numbers, including decimals. This is when an amount is rounded up or down to make it easier to work with and then adjusted at the end. E.g. add 0.9 (by adding 1 then subtracting 0.1 etc). In Year Six, children will be expected to add and subtract 1.9, 2.9, 3.9 etc and also 1.1, 2.1, 3.1 etc using this method.

- Use the relationship between addition and subtraction. State 3 facts related to one other. E.g. 1.58+4.97=6.55 so 6.55-4.97=1.58. Because knowing one fact means three others are also known, this can be used to derive new facts. For example, use 8036-1275=6761 to work out 8036-6760 or 8036-1270.

- Recognise that when numbers similar in size are added, it may be quicker to multiply e.g. when adding sets such as 70+71+75+77, realise that it is the same as (70x4) + (1+5+7). Explain the strategy.

- Check their work by comparing to their original approximation, performing an equivalent calculation, adding in a different order, using knowledge of sums or products of odd and even numbers, tests of divisibility or the inverse operation.

The following are examples of the types of calculations that children in Year Six should solve orally and/or mentally:

1. Add or subtract 4 digit multiples of 100

E.g. 5700+2500; 6200-3800; 6800-__=3700.

A range of strategies, including partitioning, compensation or using awareness of pattern in number could be used.

2. Find what to add to decimals with units, tenths and hundredths to make the next higher whole number or tenth

E.g. 6.45+_=7;   2.78+__=2.8.

Counting up and bridging through to the next tenth or whole number would be the appropriate strategies to use here. A secure knowledge of place value is required in order to use known facts in order to bridge in the context of decimals.

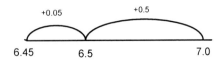

Figure 8.4

---

[29] See 'Subtraction' on page 50.

3.Add or subtract a pair of decimal fractions less than 1 and with up to 2 decimal places

E.g. 0.05+0.3   0.7-0.26.

Using awareness of pattern in number and their understanding of place value would be useful in order to solve the first example. The second example could also use these strategies but compensation or counting up and using 'bridging' could also be used. 0.7-0.26 could be seen as 0.7-0.3=0.4 then adjusted by adding 0.04; 0.4+0.04=0.44.

All of the above calculations can be solved in a range of ways and children should by now be able to choose the most appropriate method. As well as solving the examples shown above, children in Year Six should obviously also be able to solve the examples given in Year Five. It is worth referring back to this chapter in order to ensure that your child is also confident with these.

Although the above examples show what children should be able to solve mentally, it is also important for children to know when mental methods are not adequate and written methods or a calculator may be required. They need to know when it is appropriate to use a calculator, what to enter in given cases and how to interpret results (for example, know what 0.2 would mean in the context of money or length). Children working within level 5 would be expected to use the memory, bracket keys and square root keys as part of multi-step operations.

Written Methods of Recording

Children will now be expected to add and subtract decimals to 3 places as well as a range of other integers and consistent accuracy is expected in written methods. Many children may be using the standard compact methods to add and subtract. However, as long as the method in use is efficient and used with understanding, children who need to may continue to use expanded methods to solve calculations that cannot be solved mentally. Although the standard method is the eventual goal because it is the quickest, most efficient method, this is not the actual expectation until children are working at level 5 of the National Curriculum. It is better for a child to use expanded methods with understanding and accuracy than to use the standard method without a real understanding of what they are doing.

As in other years, to gain an idea of the size of the answer children should always approximate first. They should also consider the efficiency of their written method, always aiming for the most efficient method that they can use with understanding.

Developing Written Methods (Addition)

Partitioning

If your child is using an expanded method for addition, it is likely to be the method shown below which is the more efficient method. (However, other expanded methods for addition may be used and these can be found in the relevant sections in previous chapters).

```
  3   8   5   7
  2   7   7   8   +
  _____
  5   0   0   0   (add thousands)
  1   5   0   0   (add hundreds)
      1   2   0   (add tens)
          1   5   (add ones)
  _____
  6   6   3   5   (TOTAL)
```

Figure 8.5

143

Compact Written Methods

The standard method begins by adding the units and is the one with which many of us will be familiar from our schooldays. Children using it now should avoid the errors that used to occur[30] when it was taught too early, as they will have a true understanding of what they are doing. The discussion that accompanies the working out is important; for example, when adding the tens, children should be thinking 40+80 rather than 4+8. This helps to prevent errors as children are constantly thinking of the size of the numbers involved. The answer should also be approximated before beginning, in order to know if the final answer is likely to be correct.

```
    7   6   4   8
    4   4   8   6   +
  _____
  1   2   1   3   4
      1   1   1
```

Figure 8.6

With all of these strategies, children should extend their understanding so that they can solve calculations involving larger numbers, perhaps with 4, 5 or 6 digits. They should also be able to add several numbers each with a different number of digits. Should your child struggle with the compact method, then they should return to the use of expanded methods which help develop understanding.

Developing Written Methods (Subtraction)

Counting Up

This builds on the mental strategy and is discussed in more detail in previous chapters. If children are using this method, they will be encouraged to reduce the number of steps they record by solving more of the calculation mentally in order to work more efficiently.

```
    6   4   6   7
    2   6   8   4   -
      _____
            1   6   (to reach 2700)
+       3   0   0   (to reach 3000)
      3   4   6   7   (to reach 6467)
    _____
    3   7   8   3
```

Figure 8.7

Compensation

It is unlikely that your child will use this often as a written method to record.[31]

---

[30] See page 73 for a more detailed explanation.

[31] See previous chapters for further details.

```
6   4   6   7
2   6   8   4   -
3   4   6   7   (Subtract 3000)
        3   1   6   (add  316)
3   7   8   3
```

Figure 8.8

Partitioning (Decomposition)

Decomposition is discussed in more detail in the relevant sections in previous chapters. It is hoped that Year Six children using this expanded method will be able to reduce the number of steps in a calculation by partitioning and exchanging numbers mentally.

```
6 7 2          5 0 0 +  1 6 0 +  1 2
3 8 5  -       3 0 0 +    8 0 +   5  -

               2 0 0 +    8 0 +   7   =  2 8 7
```

Figure 8.9

Compact Written Method

```
  5   ¹3  1
  6̶   4̶   6   7
  2   6   8   4   -
  3   7   8   3
```

Figure 8.10

This is the most compact form of the expanded method shown in previous years which involved partitioning numbers in different ways in order to have enough from which to subtract (i.e. 'borrowing'). Even though the standard method looks identical to what was taught in schools prior to the Numeracy Strategy, the accompanying explanations should be very different with tens being referred to as tens etc. For example, in the calculation above the hundreds would be referred to as 400·600 rather than 4·6. Then when one thousand is moved or 'borrowed' the 400 becomes 1400 not 14. This helps children to retain a sense of the size of the numbers being used and avoids some of the errors that used to occur with these methods.

Children should be able to extend their understanding to work with a range of whole numbers and decimals to 3 decimal places. They should also be able to subtract numbers with varying numbers of digits.

Multiplication

It is important to understand that:

- Multiplying by one leaves a number unchanged.
- Multiplying by zero=0.

145

- Multiplication is actually repeated addition e.g. 72x3=72+72+72 and is equivalent to 72 sets of 3.
- Multiplication can be done in any order, so 56x87=87x56.
- Partitioning can be useful to help solve multiplication problems as well as addition and subtraction e.g. 12.8x60=12.8x(10x6) or (12.8x10)x6.
- Compensation can be useful to help solve multiplication problems as well as addition and subtraction e.g. 32x97=32x100-(32x3)=3200-96=3104.
- Multiplication is the inverse of division (it reverses division) and use this to check work.
- Brackets determine the order in which a calculation must be carried out.

Children must be able to:

- Count on in steps of any size (from 1 to 9) up to 100 and back from any number.
- Count on in steps of 15, 25 etc (by adding tens then the five), and steps of 9, 19, 29 (by adding multiples of ten and taking one off) and 11, 21, 31 etc (by adding multiples of ten then adding one on).
- Go beyond zero when counting back.
- Count in steps of 0.1, 0.01 and 0.001.

Children will continue the work carried out in previous years on number sequences. They should look for and explain patterns in number sequences and use this understanding to continue sequences, fill in the missing numbers in number sequences and explain why certain patterns occur. For example, 'Continue this sequence: 1, 3, 7, 15, 31, __, __.' Children should be able to explain that the rule for this sequence is to double the number and add one. Number sequences involving decimal numbers will also be explored.

Understanding of multiplication and division by 10, 100 and 1000 should be consolidated; children should have a secure understanding of the effect on the place value of the digits in a number when multiplied or divided by 10, including numbers with one decimal place, i.e. that it causes them to move 1, 2 or 3 places to the left (multiplication) or right (division).
Their understanding of multiplication and the relationship between numbers should be such that they realise that multiplication by 1000 is the same as multiplying by 10 then 10 then 10 again or by 10 then 100.

Division

Children need to continue to understand that:
- Division can be seen as both sharing equally or as grouping (repeated subtraction).
- Sharing can be more appropriate for working with smaller numbers but grouping is better for working with larger numbers.
- Dividing by one leaves a number unchanged.
- A number cannot be divided by 0.
- Division is the inverse of multiplication and should use this knowledge to check calculations.
- Division relates to fractions (i.e. know that 1/3 of 24=24/3 or 24 divided by 3).
- Division cannot be done both ways (e.g. 35 divided by 5 is not the same as 5 divided by 35).

They must also be able to:

- Give whole number remainders and decide whether to round up or down when a remainder occurs in a problem.

By Year Six, children should be familiar with the rules of divisibility and use them in their work, for example, to say whether a 3 digit number is divisible by a single digit number.

In addition to this, they should be confident when dividing decimals (e.g. 4.5 divided by 9) and should be able to relate fractions to division (e.g. 1/9 of 54=54 divided by 9 or 54/9; 23 divided by 9=2 and 5/9).

When dividing a whole number, they should be able to give the answer (quotient) as a decimal fraction e.g. 676 divided by 8=84.5, rounding to 1 decimal place if necessary. They should also be able to interpret quotients on calculators (e.g. halves, quarters, tenths, hundredths, thirds, ninths etc) as decimal fractions. Multi-stage operations should be carried out, for example, (345-66)/236+27, and decimals should be rounded to the nearest whole number or tenth in answers.

Properties of Numbers

To gain an appreciation of mathematics and the rules by which it is governed, it is important to become familiar with numbers which have specific properties or patterns. This knowledge can give children a deeper understanding of number and can allow them to use a variety of methods to calculate or check their work.

Using Tables Facts and Rules of Divisibility

Make sure your child can recognise multiples to at least 10x10 and factors of these numbers, i.e. they need to have instant recall of times tables facts and the corresponding division facts. They should also be able to find the smallest common multiple of 2 numbers, such as 8 and 12 (lowest common multiple 24). Their understanding of place value should allow them to use their multiplication and division facts to work with decimal numbers, for example, use the facts from the 6 times table to work out that 4.2 divided by 6 equals 0.7.

Patterns of multiples should be investigated and rules found. Your child will be more likely to remember a rule if they discover it through their work so it is a good idea to set your child investigations instead of simply telling them the rule. For example, ask them to prove whether the following statement is true or false: 'A number is divisible by 3 if the sum of its digits total 3, 6 or 9'. See if they can find a number to disprove the statement.

Children need to become aware of and be able to use the following rules:

- A number is divisible by 3 if the sum of the digits is divisible by 3.
- A number is divisible by 6 if it is even and divisible by 3.
- A number is divisible by 8 if it is divisible by 4 when halved or if last 3 digits are divisible by 8.
- A number is divisible by 9 if the sum of the digits is divisible by 9.
- A number is divisible by 25 if the last 2 digits are 00, 25, 50 or 75.

There are numerous rules which can be applied to help children when using multiplication and division and the list may seem rather daunting. However, many of them are actually linked and this makes remembering them much easier. For example, if children investigate the pattern of multiples of 3 and find a rule, they can use this to help them find the rule for multiples of 6. If your child does not seem aware of all of them, don't panic! Simply encourage them to be aware of and explore the relationships between multiples and their factors so that they can use those that they do know to work more flexibly and efficiently.

Using Factors.

A factor of a number will divide into that number. For example, the factors of 6 are 1, 2, 3 and 6 since 6 is divisible by them all. Once children have an understanding of factors and how they can be used, they can make certain calculations easier to solve. Help your child to become aware of how the following can help them to work with increased flexibility.

- Use factors to find products mentally or with jottings. So to solve 42x18 mentally: 42x2x9=84x9=84x3x3=252x3=756.
- Identify numbers with an odd number of factors (square numbers). Because a square number is multiplied by itself (e.g. 16 is a square number because 4x4=16), when all the factors are found, there will be an odd number. All the factors of 16, for example, are 1, 16, 2, 8, 4. This can be set as a challenge for your child. 'True or false: all numbers have an even number of factors?'
- Identify 2 digit prime numbers (i.e. numbers with only 2 factors). An example of a prime number is 17 as the only factors of 17 are 1 and 17.
- Know instantly prime numbers up to 20.
- Factors are governed by certain rules and patterns. For example, because 32 is a multiple of 8, it is also a multiple of all the factors of 8 (i.e. it must be a multiple of 1, 8, 2 and 4).

Square Numbers

These are numbers which are multiplied by themselves e.g. 4x4; 9x9. Your child should be able to quickly work out square numbers up to 12x12 and use this to work out the corresponding squares of multiples of 10. For example, if they know that 5 squared is 25 then they should be able to work out that 50 squared would be 2500. They should be given the opportunity to explore square numbers so that they can practise finding the squares of larger numbers (e.g. 15 squared, 21 squared) and can learn how to identify 2 digit numbers which are the total of 2 square numbers, e.g. 61 =5 squared + 6 squared. Problems may be given which involve finding out. For example, 'Which number when squared totals 4225?'

In the following problem children would be expected to approximate an answer and then use this as a starting point for using a calculator to find the actual answer. 'The area of a square wall is 4624. What length must the sides of the wall be?' Your child should be encouraged to use their tables knowledge to approximate first, so 60x60=3600 and 70x70=4900 so the correct answer must lie between 60 and 70. They could then use their calculator to find the actual number. Through appropriate questioning, you can encourage them to think about the best place to start. Is 4624 nearer to 3600 or 4900? So is it a good idea to begin by trying 61 squared or would it be wiser to start with a number nearer 70 squared? Exercises such as this can help your child to see the value of approximating and thinking logically before trying out possibilities. Children working at level 5 of the National Curriculum are also expected to recognise the square roots of perfect squares up to 12x12.

Prime Numbers and Prime Factors

A prime number is a number which has only 2 factors (1 and itself) so, for example, 7 is a prime number as it only has the factors 1 and 7. In Year Six, your child will be expected to be able to identify all the prime numbers up to 100 and instantly spot those below 20.

Prime factors are the lowest factors of a number that can be multiplied together to make that number. For example, the prime factors of 42 are 2, 3 and 7 since 42=2x21 =2x (3x7). Children should be able to find all the prime factors of any number to 100 by finding a pair of factors and then reducing each factor into its factors until only prime numbers remain. To find the prime factors of 36, for example, the process could be: 6x6=(2x3)x(2x3) or 2x18=2x (2x9)=2x2x(3x3). The prime factors are 2, 2, 3 and 3.

This sort of work can help to deepen understanding of multiplication and give children ideas about new ways to work.

Rapid Recall (Multiplication and Division)

Know by heart:

- All tables facts up to 10x10 (including the effect of multiplying by 0 and 1) and corresponding division facts.
- Know squares of all numbers up to 12x12.

Derive quickly:
- Related facts such as 6x7=42 so 600x7=4200.
- Squares of multiples of 10 to 100 (e.g. 20 squared).
- Doubles of 2 digit whole numbers or 2 digit decimals (e.g. double 36=72; double 3.6=7.2) and corresponding halves.
- Doubles of multiples of 10 to 1000 and corresponding halves.
- Doubles of multiples of 100 up to 10,000 and corresponding halves.
- Continue to be aware of the effect of multiplying or dividing decimals by 10 or 100 and whole numbers (integers) by 1000.
- Use known multiplication and division facts to work out calculations involving decimals e.g. 4x0.8=3.2; 7.2 divided by 9= 0.8.

It is important to have a secure understanding of the inverse relationship between doubling and halving (i.e. one reverses the other) and be able to use this knowledge to work more efficiently during calculations. This understanding should extend to working with decimals e.g. double 5.6=11.2; half 11.2=5.6.

Mental Calculation Strategies (Multiplication and Division)

Children should know and be able to use the following:
- Use partitioning to find doubles and halves, e.g. double 456; double 400, double 50 and double 6 then add the total.

- Multiply a 2 digit number by a single digit mentally by partitioning. E.g. 47x4= (40x4)+(7x4).

- Multiply 2 digit decimals by a single digit mentally by partitioning. E.g. 8.6x7=(8x7)+(0.6x7)=56+4.2=60.2

- Use compensation, that is round numbers up/down to make them easier to work with then adjust, e.g. 56x90; (56x100)-(56x10). To multiply a number by 49 or 51, multiply it by 50 then subtract or add the number. To multiply a number by 39 or 41, multiply it by 40 then subtract or add the number. Again, to multiply a number by 99 or 101, use the same strategy.

It is important to be aware of pattern and relationships between numbers as this can help children to find alternative ways of working if they need to.

- For example, know that by doubling/halving a factor in a multiplication, the answer will be double/half. An example of this would be 7x4=28 so to find 7x8 you can simply double 28. This type of strategy can be used if a table fact is not known or to check work using an alternative method.

- Be aware of the relationship between numbers when they are multiplied; the same product can be reached by halving one factor in an equation and doubling the other. E.g. 6x5=3x10 or 5x24=10x12. Children in Year Six should be able to solve such calculations by doubling a factor ending in 5 and halving the other.

- Know that new facts can be found by combining known facts. For example, to find 17x7 the 10 times table and the 7 times table can be used: (10x7) + (7x7) =70+49=119 so 17x7=119.

- Know that to multiply by 15, multiply by 10 then halve it and add this on; or multiply by 30 then halve.

- Know that to multiply by 25, multiply by 100 then divide by 4 (halve it and halve it again).

- To find facts for the 24 times table, find the facts for the 6 times table then double it and double it again.

- Use combinations of facts to find multiples of 32. For example, 2x32=64 so double this to find 4x32(=128), double it again to find 8x32(=256), double it again to find 16x32(=512). Work out other multiples by combining facts. E.g.12x32= (8x32) + (4x32).

- To find 1/6, find 1/3 then halve it; find 1/12 by halving a 1/6; find 1/20 by halving a tenth.

- Use knowledge of factors e.g. 43x24=43x(3x8) so 43x3=129; 129x8= 129x2x2x2=1032. This also works for division e.g. 180 divided by 30 could be solved as 180/10=18 then divided by 3=6.

- Use knowledge of the inverse relationship between multiplication and division with whole numbers and decimals. Knowledge of fractions and their relationship to multiplication and division is included here. E.g. 5x60=300 so 1/5 of 300=60 and 1/6 of 300=50. ¾ of 4=3 so 4x ¾=3.

- Use approximations and tests of divisibility to check results.

Again, although this is rather a long list, most of the facts simply require a secure understanding of the concepts of multiplication and division, an appreciation of the relationships between numbers (particularly doubles and halves) and the willingness to work flexibly. By knowing how to use numbers flexibly, children can solve calculations in a variety of ways, thus drawing on their own strengths and preferences when working.

It is expected that children in Year Six should be able to work mentally to multiply or divide a 2 digit number (or a decimal to one place) by a single digit and to solve calculations such as the following.

1. Multiply a decimal fraction with one or two decimal places by 10 or 100

E.g. 32.45x10=324.5

2. Divide a one or two digit whole number by 100 or 10

E.g. 4.6 divided by 100=0.046

3. Double a decimal fraction less than 1 with one or two decimal places

E.g. Double 0.75=1.5.

Awareness of pattern in number is the most useful strategy here.
75x2=150 so 0.75x2=1.5.

4. Halve a decimal fraction less than one with one or two decimal places

E.g. 0.42 divided by 2=0.21

Use awareness of pattern in number: half of 42 is 21 so half of 0.42 is 0.21.

5. Multiply a decimal fraction by a single digit number

E.g. 0.8x9=7.2

Again awareness of pattern in number and understanding of place value is needed.

6. Multiply a 2 digit whole number or decimal fraction by any single digit number

E.g. 48x7 or 4.8x7.

Partitioning these numbers mentally would probably be useful here.

7. Divide a 2 digit whole number or decimal fraction (to one place) by any single digit number

E.g. 45 divided by 8; 6.4 divided by 8.

Use of tables facts and (in the case of decimals) knowledge of place value is required here.

Developing Written Methods (Multiplication)

It is expected that most children will be using an efficient written method confidently to multiply 3 digit numbers by a single digit, 2 and 3 digit whole numbers by 2 digit numbers and decimals by a single digit. For some children this will be the standard compact method. However, this will not necessarily be the case. As long as the method used is efficient and the child has a good understanding of what they are doing, expanded methods are perfectly acceptable.

Grid Method

This can be used with numbers of any size, including decimal numbers. It involves partitioning numbers in much the same way as children are encouraged to do during mental methods. It is more efficient if the numbers are added together mentally to cut down on the amount of recording. Children are expected to multiply 4 digit numbers by a single digit and 3 digit numbers by 2 digit numbers.

Short Multiplication.

E.g. 5378x6

Approximately it would be 5000x6=30000

x	5000	300	70	8
6	30000	1800	420	48

Figure 8.11

Total: 30000+1800+420+48=32268

7.34x8

x	7	0.3	0.04
8	56	2.4	0.32

Figure 8.12
Total 56+2.4+0.32=58.72.

Long Multiplication

Even if children are confident using the compact method for multiplying 2 digit numbers, it is likely they will revert to the grid method at least briefly when learning to multiply 3 digits by 2 digits.

756x26

x	7 0 0	5 0	6	
2 0	1 4 0 0 0	1 0 0 0	1 2 0	=15,120
6	4 2 0 0	3 0 0	3 6	=4536

Figure 8.13

Total: 15,120 + 4536 = 19,656

The revised strategy has advised that the grid could be presented with the larger number at the side and the smaller number across the top (examples are given in Year Five) in order to make the addition of the partial products easier. If supporting your child with the grid method, it would therefore be wise to find out which way their school presents the grid so that you can work in the same way.

Expanded Method

Short Multiplication

```
    5  3  7  8
             6  x
  _____
  3  0  0  0  0  (6x5000)
     1  8  0  0  (6x300)
        4  2  0  (6x70)
        4  8  (6x8)
  _____
  3  2  2  6  8
  _____
```

Figure 8.14

Eventually leading to the more compact form:

152

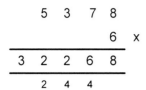

```
    5   3   7   8
                6   x
  ─────────────────
  3   2   2   6   8
      2   4   4
```

Figure 8.15

Long Multiplication

The revised strategy states that most children should aim to be confident with the more compact method for multiplying two 2 digit numbers (shown below) by the end of Year Five. However, efficient methods that are not necessarily the standard method are perfectly acceptable. (Examples of these are shown in the relevant section in Year Five). It is not until children are working at level 5 of the National Curriculum that standard methods are actually expected. It is worth mentioning that, in the example below, the carry digits in the partial products of 56x20=1120 and 56×7=39 are usually carried mentally.

```
      5 6
      2 7  x
  ───────────
  1 1 2 0  (56x20)
    3 9 2  (56x7)
  ───────────
  1 5 1 2  (Total)
      1
```

Figure 8.16

The expanded method for multiplying a 3 digit number by a 2 digit number is extremely cumbersome and may be useful to discuss with regard to its efficiency. It can be shown as a link between the grid method and the standard compact method but it is advisable to move onto the compact method in as little time as possible.

```
          7   5   6
          2   6   x
  ─────────────────────
      1   4   0   0   0
          1   0   0   0
              1   2   0   +
          4   2   0   0
              3   0   0
                  3   6
  ─────────────────────
      1   9   6   5   6
```

Figure 8.17

Standard Compact Method

As in previous examples, the carry digits in the partial products are usually carried mentally. These methods should also be used with decimals with up to 2 decimal places.

```
        7   5   6
            2   6   x
  ─────────────────
1   5   1   2   0   (756x20)
        4   5   3   6   (756x6)
  ─────────────────
1   9   6   5   6
  ─────────────────
```

Figure 8.18

Developing Written Methods (Division)

Again, standard methods may be in use but this will not necessarily be the case. Many children will still be using expanded methods. It is better to use expanded methods with accuracy and understanding than the standard method without a full understanding. However, by this stage when expanded methods are in use it is important to encourage your child to work as efficiently as possible, only recording stages that cannot be carried out mentally. In Year Six, children will be expected to divide 2 and 3 digit numbers and decimals by a single digit.

Using Multiples of the Divisor

As in previous years, this relies upon repeated subtraction and using tables facts to subtract large groups of the divisor. The question to begin with is: 'What is the highest multiple of ten which the divisor (in this case 6) will divide into?' Children often begin by making use of their ten times table. When working with larger numbers they can be encouraged to use their ten times table in order to find a starting point at which to subtract. In the example below, to decide what to subtract initially, children can work out that because 10x6=60, 20x6=120. This still seems relatively low in comparison to 455 so they can be encouraged to work out 30x6 (usually by saying 3x6=18 so 30x6=180). This is reasonably large and will not make the calculation too long winded. It would not be wrong to subtract 20x6; it would just be less efficient and more time consuming. With practice, children become more used to finding appropriate multiples to subtract.
3 digit divided by 2 digit number: 455 divided by 6

```
        4   5   5
    -   1   8   0   (6x30)
      ─────────
        2   7   5
    -   1   8   0   (6x30)
      ─────────
            9   5
    -       6   0   (6x10)
          ─────
            3   5
    -       3   0   (6x5)
          ─────
                5
```

Figure 8.19

So the answer is 75 remainder 5 (30+30+10+5), as 75 groups of 6 were subtracted from 455, with 5 remaining.

Once children increase their understanding of the method, and their mental skills alongside this, a more efficient method for short division of HTU by U can be done through partitioning rather than repeated subtraction (chunking).

Partitioning.

455 divided by 6
Begin by approximating: 6x70=420 and 6x80=480 so the answer will between 70 and 80. This also gives a starting point from which to start partitioning.

$$\frac{70 + 5 \; \text{R5}}{6 \overline{)420 + 35}} \quad = 75\text{R5}$$

Figure 8.20

This can then be shortened to the compact written method for dividing 3 digit numbers by a single digit:

$$\frac{7 \quad 5 \; \text{R5}}{6 \overline{)45 \;{}^3 5}}$$

Figure 8.21

The accompanying discussion needs to be based around the real numbers so instead of saying 'How many sixes are in 45, the question is 'How many sixes are in 450?' Or 'What is the highest multiple of ten that 6 will divide into?' The six times table can be used to derive new facts, 6x7=42 so 6x70=420.

Long Division

This is actually now included in the Year Six section of the Primary Framework entitled 'Progression to Year Seven' as it requires extremely secure mental skills. It would usually be introduced in Year Six as it is appropriate for children working within level 5 of the National Curriculum. Long division is used to divide by larger numbers, for example, a 3 digit number divided by a 2 digit number. The chunking method (repeated subtraction) used in previous years is used here.

For example, 639 divided by 45. Begin by using the ten times table as a starting point both for approximation and for knowing what to subtract first. 45x10=450 and 45x20=900, so the answer must be between 10 and 20. Subtract 450 first which leaves 189. Doubling and doubling again can be used to find that 45x4=180.

```
              1   4   R9
        _____
   45 ) 6   3   9
      -   4   5   0   (45x10)
        _____
          1   8   9
      -   1   8   0   (45x4)
        _____
                  9
```

Figure 8.22

The answer is 14 remainder 9 as 45 was subtracted 14 times from 639. This is essentially the same process as used for traditional long division although it is laid out slightly differently. The questions asked during the course of their work are also phrased differently so that the emphasis remains on mental methods.

<u>Checking Work</u>

To improve accuracy and awareness of relationships between numbers, children are required to check their work as an integral part of their working out. In earlier years, children will be encouraged to do this, but it does become much higher profile as the children become more mature and as the complexity of the calculations increase. Each calculation should begin with an approximation of the answer, made by rounding the numbers involved up or down. For example, an approximation of 45x33 could be 50x30. The children would know therefore before they started the calculation that the answer should be in the region of 1500. There are also a number of other strategies which are encouraged. Children need to be taught them all so that they can decide which is the most appropriate to use at the given time. The following are the methods with which children will be expected to check their work:

- Check using the inverse operation, i.e. use subtraction to check addition and vice versa; use multiplication to check division and vice versa; use doubles to check halves and vice versa. For example, 321-176=145. 145+176=321. 320 divided by 4=80. 80x4=320.
- When adding three or more numbers, add in reverse order, e.g. 375+238+242=855. 242+238+375=855.
- Carry out an equivalent calculation (i.e. work it out in a different way). For example, 49x6= (40x6) + (9x6) or (50x6)-6.
- Approximate the answer before calculating so that the child has a rough idea of what the size of the answer should be.
- Use knowledge of sums or products of odd or even numbers.

To summarise, it is expected in Year Six that children will have a good grasp of number relationships and patterns as well as a secure knowledge of facts requiring instant recall. These will help them to work with confidence, speed and flexibility when solving problems. If they are able to bring facts to mind quickly when working, it removes the problems caused by having to laboriously count and work things out. Encourage your child to use what they know to calculate in daily life. Remind them of strategies that will help to simplify calculations and practise using them whenever possible in real life contexts.

More confident children, who are likely to be working within level 5 of the National Curriculum, may also begin to use their knowledge of multiples, factors, divisors and common factors in simple cases. However, all children should continue to use the methods and strategies taught in previous years and also explore new rules and patterns. Remember to ensure that you give your child the opportunity to find out and discover these rules and patterns themselves because, as well as being more interesting, it also helps them to remember what they have learnt. Experimenting with number can also increase their confidence and motivation in mathematics and will help to prepare them for the secondary school curriculum.

Chapter 9 :   Games to Play

In this chapter I have tried to give some simple ideas for games to play which will develop your child's understanding of number order and relationships. Most can be adapted for use with any year group so, for example, a Reception child could use them to develop an understanding of number to 10 whilst a Year Six child could use them to develop an understanding of decimals with three decimal places. They can be easily made using pieces of card and, if you have access to a laminator (small ones can be bought from most stationers and are quite reasonably priced), resources that will be well used can be laminated to make them last longer. These types of resources are also available to download from my website www.mathematicsathome.co.uk.

Digit Cards

Put out cards with the digits 0 to 9 on them and ask your child to put them in the correct order. Then give instructions such as, 'Cover the number before 5', 'Turn over the number in between 8 and 10', 'Point to the number that is one less than 7' etc. Every time the child follows the instruction correctly, they get a point/smiley face/sticker and if they get it wrong you get the point (or you could take a turn as long as you make sure that you don't always win!!) The first person to reach 5 points is the winner. This game can be made harder by taking it in turns to gives each other instructions. Even if the instructions your child gives are not quite right, try to follow them or rephrase them in a way that you can follow so that your child gets the chance to feel important and on an equal level with you. These games are designed to develop confidence so make light of any mistakes in order to keep your child motivated. Obviously, by giving you the instructions they are using the language themselves and thus deepening their own understanding. You could check the answers together each time by counting along the number track or line of numbers.

This can also be played as an outdoor game; numbers can be chalked on the ground to make a number track (see Figure 9.1) or stepping stones can be made into a number track then the child can be asked to jump on the number before 4, for example. You could then check together by counting along and award your child the point. If they are not correct, remember to give clues and give your child a chance to put it right before awarding the point. By correcting themselves, children often learn not to make the same mistake next time.

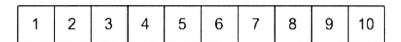

Figure 9.1

To adapt the game for older children, try ordering teens numbers or any group of 2 or 3 digit numbers. Digit cards can have any number written upon them. Older children could try ordering decimals, fractions, negative numbers or certain multiples of a given number. Consecutive numbers do not necessarily have to be used. A group of random numbers can be ordered and your child could identify numbers before, after or between others. Cards can also be sorted. For example, fraction cards could be sorted into those more or less than a half.

Make the Number

A similar game is to make (or download) some digit cards with the numbers 1 to 9 on them and ask your child to make the largest or smallest number possible with them. For example, you could give 5 cards and ask them to make the smallest 2 digit number or the largest 3 digit number they can. Instructions such as 'Make me a number with 5 tens' or 'Make a number that is between 345 and 365' can be given. A decimal point could also be included for older children. Children quite enjoy the challenge of being asked if they can find all the numbers that can be possibly made with the cards or all the numbers over, for example, 500. As your child is working ask them questions such as 'How can you be sure that is the largest number you can make?' This helps them to explain and clarify their thinking and gives you an indication of their understanding.

Hide the Number

This game involves covering a number on a number track for your child to guess which one is hidden, explaining why. For example, you could have the numbers from 1 to 10 on a number track or numberline. The child closes their eyes then you cover a number. When they open their eyes, they have to identify the number which is hiding. Then you close your eyes and they hide a number from you. Obviously a point is gained if the number is guessed correctly. If the winner is the first person to reach 5 points, then make sure that your child has the first turn so that they win! The idea of these games is to build their confidence not to explore the concept of 'life isn't all about winning!'

As part of this game, it is important to encourage your child to explain their thoughts. For example, 'That's clever. How did you work that out?' Much of the learning from most of these games will come from the discussion of how the child knows if they are right. For example, if 4 is hidden, the child could explain that they knew it would be 4 because it was in between 3 and 5, or because it was the number after 3 or one less than 5. Make sure that you give explanations when it is your turn as it will model the type of language that you want your child to use. Before, after, more, less, greater, smaller, in between are all words and phrases with which young children will need to be familiar.

To use this game with older children, try increasing the size of the numbers, using decimals, negative numbers or multiples of a given number. For example, write all the multiples of seven on cards and put them in order. Hide a multiple and ask your child to identify the number hidden or swap two multiples over and see if your child can identify which two have been swapped. Again encourage them to explain their thinking and take a turn yourself so you can model the language if your child finds it difficult. For example, 'I knew that card would be 21 because it is 7 more than 14'. Relationships between doubles and halves can also be discussed. 'I knew 4x7 would be 28 because I doubled 14 (2x8)'.

Place the Number

In this game number cards from 1 to 10 are spread out face down on the table and a blank number track (or a number track with just a few numbers on it, if you feel this is more suitable for your child) is displayed.

Figure 9.2

The players take turns to pick a card and to position it in the correct place on the blank numberline, explaining their thinking as they do it. This game is useful to help demonstrate the range of methods there can be for working something out. For example, when the first number is picked, it is likely that you will need to count to find its position as there will be no other numbers on the number track to help you. If the first number picked was 9, for example, the point can be made that it is actually easier to count back from 10 than on from 1 when placing it. We know that 9 is one less than 10 and the number track goes up to 10 therefore we can use this to place 9. As more numbers are added to the numberline, the game becomes easier in terms of placing the numbers. However, the explanations remain useful. For example, if 5 is picked and the number 3 is already on the number track, instead of counting from 1 to find 5, you can simply count up from 3. When playing this game, it is important to demonstrate counting back from numbers where appropriate as well as counting forwards. Your turns are as important as your child's as they demonstrate to your child alternative strategies for finding numbers. You can make your own rules as to the winner of the game. It can be the person who place the most numbers or the person who reaches 5 points first. This game can also be varied to make it shorter by only giving out 5 cards to each player and saying that the first person to place all their cards is the winner.

When using larger numbers, this game can be used to develop an understanding of place value. If you make some place value arrows, then your child could pick a digit card with, for example, 45 on it and make 45 using the place value arrows. If they make it correctly, they can place the digit card (or write the number) in the correct place on a numberline and gain a point. The person who places the most cards correctly is the winner. If you have difficulty making a numberline long enough for larger numbers, then use a blank numberline and label it differently. You could start the numberline at 456 and end it at 466, or 556, depending on what suits your purposes.

Multiples of a certain number can be positioned on a numberline in order to improve understanding of times tables. So all the multiples of 8 could be written on cards and then you could place 8 at the start and 80 at the end of the numberline and ask your child to identify the multiple to go in a given position. You could ask, for example, the position of 72 and then get your child to explain their thinking. They may, for example, say I knew 10x8 was 80 so I subtracted 8 from 80 to find what 9x8 would be. Questions can bring attention to number relationships. For example, you could ask your child to identify the number which equals 2x7 then ask them to use this knowledge to predict 4x7 (by doubling).

Guess the Number

This game involves players thinking of a number and writing it on a piece of paper. The object of the game is to guess the number chosen by the other player by asking questions. A numberline is needed for each player in order to eliminate numbers. For example, if my number was 46, the other player could ask questions such as 'Is it greater than 50?' The answer would be no so the numbers over 50 could be marked off on their numberline. On the next turn the question could be 'Is it an even number?' or 'Is it a multiple of 10?' Gradually more numbers can be eliminated until the actual number is found. Rules can be made such as you must have a guess after asking five questions. This game can easily be adapted to for different age groups by varying the size of the numbers or extending into decimals and using more complex mathematical language.

Pairs

There are numerous variations of this game. It can be played to develop rapid recall of number bonds by having cards with dots or numerals drawn that total a certain number. For example, the object of the game can be to pick pairs that total 10 so when a player turns over a card with 6 on it, they need to work out that they need the card with 4 in order to win the pair. Numbers can be varied to make it more challenging for older children. For example, children could find multiples of ten that total 100 or decimals (tenths) that total 1.

Another way to play is to make the object of the game to add or multiply the two numbers that are turned over in order to win the pair. Obviously, the player with the most pairs at the end is the winner.

Dice Games

There are a number of these games available to download from my website. All that is required is a dice and some counters or coins. Each player has counters of a given colour (so, for example, player one may be red and player two blue) and a playing board with various randomly arranged numbers on it. If the game is being used to develop knowledge of doubles and halves then the numbers on the playing board would be double those on the dice

(Remember blank dice can be bought so that you can write your own numbers on if you wish). Once the dice is thrown the player has to double the number thrown and cover up that number on the playing board with one of their counters. This game is good for developing an understanding of inverse relationships as once there is only one number left on the board (for example, 12) the players have to work out what they need to throw (in this case 6) in order to cover the number. Once all the numbers are covered, the person who has covered the most numbers with their counters is the winner.

The same game can be used to develop knowledge of number bonds. The numbers on the playing board would be whatever was needed to be added to the number on the dice to make the total. So if the game was based around number bonds to 10 then the numbers would be those that can be added to 1, 2,3,4,5 or 6 to make 10.

To use this game with older children, the object could be to multiply a given number, for example 6, by the number on the dice then cover the answer on the board. Alternatively 2 dice could be rolled. Larger numbers written on blank dice could be used to, for example, divide multiples of 6 (on the dice) by 6 and cover up the answer on the board.

The possibilities for this game are endless and all that is really needed are counters, playing boards and dice. If you cannot find dice with appropriate numbers then cards with relevant numbers written on could be turned over instead of rolling a dice. I have included blank boards on my website so that you can write in your own numbers and tailor the game to your child's particular needs. Alternatively, numbers could be chalked on the ground outside and a giant dice used. If your child dislikes sitting down and 'working' then it makes sense to play outside and include plenty of opportunities to jump, hop, throw beanbags etc at different numbers.

These types of games should only last a few minutes but can be played quite often to build up your child's confidence with number relationships, number order and the concepts of counting on and counting back.

It is also essential to vary the game according to your child's stage of development. Remember that if your child finds an activity difficult, then smaller numbers should be used to build up confidence with the concept. Once the concept is grasped with smaller numbers, your child will be far more likely to understand the original activity when it is returned to.

Conclusion

The National Numeracy Strategy was introduced in order to improve the teaching of mental methods in mathematics and thus make British children more numerate. This book demonstrates the numerous strategies that can aid mental methods and gives you as a parent an awareness of what is expected in each year group.

Underpinning all of the mental strategies is a sound knowledge of important number facts and these are used repeatedly to speed up calculations. A secure understanding of place value is also vital. Once this is in place, knowledge of patterns in numbers again can be used to work more efficiently. Although written methods are used, they are introduced only when mental strategies and number facts are firmly in place.

The written methods introduced initially link to mental methods and still rely on an understanding of place value and use of known facts. It is only when children have a true understanding of the numbers with which they are working that standard written methods are introduced.

One point that I feel the Numeracy Strategy makes very well is that maths is not about applying a procedure to numbers to gain an answer; it is about using numbers logically, making use of their properties, patterns and the knowledge you already possess to solve problems. In order to become numerate, children must be able to make decisions so it is important to discuss the relative values of different strategies in different situations, bringing your child's attention to cases where one strategy may be more useful than another. Unless your child is really struggling with a particular area of mathematics and needs your help to practise examples, I would advise against sitting down and completing pages of 'sums'. It makes far more sense to encourage the use of maths in context. For example, draw their attention to the use of measuring jugs, scales and clocks in various contexts and ask them questions during the course of activities. A measuring jug can be seen as a vertical numberline so questions related to ordering and rounding numbers can be asked and worked out practically. Similarly, in shops, discuss prices and reductions and ask questions based around these occasionally to aid understanding of percentages or decimals. If written activities are given, then give them as a challenge. For example, give a statement to prove or disprove. Activities which involve making decisions and solving problems encourage children to think logically and look for patterns which in turn increases their numeracy and also, hopefully, makes maths more interesting!

Remember that this book is designed to be a reference book. Use it to gain an understanding of the level at which your child should be working and the strategies they need. Once you have this information, encourage your child to use mathematical skills wherever possible in their everyday life and help them to apply the skills they have in problem solving contexts. Mathematical skills are required for daily life and therefore the most important thing that your child can gain from their mathematical learning is to use numbers in a logical, efficient way. By supporting them with this in primary school, when they are laying the all important foundations, you will help to provide them with the confidence and skills required not only for the secondary school curriculum but also for their later life.

Glossary

Compensation- This is the strategy when numbers are rounded up or down to the nearest multiple of ten or hundred to make them easier to work with, then adjusted at the end to compensate for this rounding up/down. For example, in the early years add 9 may be solved by adding 10 and then subtracting 1, as it is easier to solve mentally. 45+39 may be seen as (45+40)-1 to make it easier to solve mentally. 45+40 is worked out then 1 is subtracted at the end. £1.99x3 is easier to solve if it is seen as (£2x3)-3p. This is an important mental strategy and often removes the need for written recording.

Factor- A factor of a number will divide into that number. For example, 6 is a factor of 12 because 6 will divide into 12. It is therefore also a factor of 24, 36 etc. The factors of 20 are 1, 2, 4,5,10 and 20 because 20 can be divided by all these whole numbers.

Fractions.
Mixed Fractions-These are numbers made up of a combination of a fraction and a whole number. For example, 4 ½ is a mixed fraction.

Unit Fractions-These are fractions that are one part of the whole (i.e. the top number is always one). For example, ½, 1/3, 1/10 etc.

Simple Fractions/ Proper Fractions-These are fractions where the top number (numerator) is smaller than the bottom number (denominator). For example, 3/5 or 5/8.

Improper Fractions-These are fractions where the numerator is larger than the denominator. For example, 10/8 or 5/2. These can be converted to mixed fractions so 10/8 would be equal to 1 and 2/8 or 1 ¼.

Hundred Square-An example of a hundred square is shown below. These can be easily made using a computer or they are available to buy as posters from many large stationers and bookshops. Hundred squares are useful as they show the patterns made by adding and subtracting tens to numbers, if read vertically as well as showing the sequence of numbers from 1 to 100.

1	2	3	4	5	6	7	8	9	10
11	12	13	14	15	16	17	18	19	20
21	22	23	24	25	26	27	28	29	30
31	32	33	34	35	36	37	38	39	40
41	42	43	44	45	46	47	48	49	50
51	52	53	54	55	56	57	58	59	60
61	62	63	64	65	66	67	68	69	70
71	72	73	74	75	76	77	78	79	80
81	82	83	84	85	86	87	88	89	90
91	92	93	94	95	96	97	98	99	100

Figure 11.1

Integer- This means a *whole* number so a number which is not a fraction or does not include decimals.

<u>Inverse</u>- The inverse operation is the opposite operation, so addition is the inverse of subtraction and vice versa; multiplication and division are the inverse of each other and doubling and halving have an inverse relationship.

<u>Mental Calculation Strategies</u>-These are the strategies which can be used when working mentally in order to solve calculations in an efficient way. Certain mental calculation strategies are taught in each year group and built upon year after year.

<u>Near Doubles</u>-These are numbers which are nearly a double and thus can be worked out using knowledge of doubles rather than other methods. So, for example, 5+6 is a near double. It is one more than 5+5 or it is one less than 6+6. Once children know this, it can be used to speed up calculations. Calculations such as 300+400 would also class as a near double, as would 300+299. By recognising that these are close to a double they can be solved without counting.

<u>Number Bonds</u>-These are the number which 'bond' to form other numbers. Number bonds to ten, for example, would mean all the numbers which add together to total ten and the subtraction facts that correspond with them. So 1+9=10, 9+1=10, 10-9=1, 10-1=9. 2+8=10, 8+2=10, 10-8=2, 10-2=8 etc.

<u>Partition(ing)</u>- This means to split numbers. It often refers to splitting numbers into thousands, hundreds tens or ones, depending on the size of the number. For example, 345 can be partitioned into 300+40+5. Partitioning numbers in this way is encouraged when adding, subtracting, multiplying and dividing as it makes numbers more manageable and easier to work with mentally.
Partitioning can also mean splitting a number in other ways. So, for example, if subtracting 7 from 65, 7 may be partitioned into 5 and 2 to make it easier to work with, as mentally it is quicker to work out 65-5-2 than 65-7.

<u>Place Value</u>-This is literally the value indicated by the place that a digit occupies. For example, 1 represents 1 unit in the number 41 but represents 10 in the number 14. Where the 1 is placed indicates its value. The term place value is used to mean tens and units or hundreds, tens and units etc.

<u>Place Value Arrows</u>-These are small cards with an arrow shape on the right. The arrow shows the hundreds, tens and the other single digit numbers.

Figure 11.2
These can help children to understand how numbers over ten are made up. In the example above, if the 8 is placed on top of the ten (with the arrows lined up) it will make the number 18. The children can see that the one is actually still a ten but with the zero hidden. This is very important as some children don't realise the significance of number order so they see 18 as 1 and 8 instead of 10 and 8. This in turn leads to confusions between 18 and 81.
These arrows can also be used to make 2, 3 or 4 digit numbers. If the 18 made in the previous example, is then placed on top of the hundred (hiding both zeros), it will make the number 118. By taking the place value arrows apart again, it is easy to see that 118 is made up of 100+10+8. This helps children to realise that the first one in 118 is actually a hundred (just with the zeros hidden). This helps them learn how to record 3 digit numbers, as many children find this confusing initially, recording 118 as 10018.

Product- This is the result when 2 or more numbers are multiplied together, e.g. the product of 4 and 5 is 20 because 4x5=20.

Quotient- This is the answer gained when one number is divided by another, so 4 would be the quotient if 20 were divided by 5.

Tens Boundary/Hundreds Boundary-This refers to when the next multiple of ten or one hundred is crossed in a calculation. Calculations which cross the ten/hundreds boundary are harder than those which do not cross. For example, 35+2 does not cross the tens boundary, but 35+7 does cross the tens boundary as it involves crossing the next multiple of ten. 304-3 does not cross the hundred boundary but 304-9 does, as it involves crossing the previous multiple of one hundred.

Index

About the Author

Michelle Cornwell graduated from Durham University in 1992 and has worked as a primary school teacher in the North East of England for the last 16 years. She became Mathematics Co-ordinator at her first school in 1993 and currently holds the post of lower school Numeracy Co-ordinator in her present school.

Acknowledgements

The author would like to thank Roma Cockett and Kate Paige, the Numeracy Consultants at Redcar and Cleveland Borough Council, for taking the time to read and comment upon this book. Their support and advice has been very much appreciated.

Lightning Source UK Ltd.
Milton Keynes UK
UKOW022134231012

201062UK00010B/1/P

9 780955 692000